Fundamentalism and Education in the Scopes Era

Fundamentalism and Education in the Scopes Era

God, Darwin, and the Roots of America's Culture Wars

Adam Laats

palgrave
macmillan

FUNDAMENTALISM AND EDUCATION IN THE SCOPES ERA
Copyright © Adam Laats, 2010.

First published in hardcover in 2010 by PALGRAVE MACMILLAN® in the United States—a division of St. Martin's Press LLC, 175 Fifth Avenue, New York, NY 10010.

Where this book is distributed in the UK, Europe and the rest of the world, this is by Palgrave Macmillan, a division of Macmillan Publishers Limited, registered in England, company number 785998, of Houndmills, Basingstoke, Hampshire RG21 6XS.

Palgrave Macmillan is the global academic imprint of the above companies and has companies and representatives throughout the world.

Palgrave® and Macmillan® are registered trademarks in the United States, the United Kingdom, Europe and other countries.

ISBN: 978-1-137-02101-4

Library of Congress Cataloging-in-Publication Data is available from the Library of Congress.

A catalogue record of the book is available from the British Library.

Design by Scribe Inc.

First PALGRAVE MACMILLAN paperback edition: August 2012

10 9 8 7 6 5 4 3 2 1

Printed in the United States of America.

Contents

Preface

The research for this book was not motivated by a desire to defend or to attack fundamentalism. I was not raised in a fundamentalist or evangelical church. My family was loosely attached to a number of different liberal Protestant churches and now I attend Catholic services. Nevertheless, as every historian must do, I have tried to develop a sympathy for the goals of fundamentalists and evangelicals in order to gain a fuller understanding of their educational campaigns. In the end, however, I take the position of an outside observer. This distance has probably made me slower to pick up some of the nuances of fundamentalist language than an insider might be. But it has also allowed me to avoid the temptation to gloss over the parts of fundamentalism that do not match my ideals. Of course, no writer can avoid coloring his research with his own perspective. However, this study has been motivated by an academic interest in a complex theological and educational movement, not by a desire to justify or condemn the activism and ideas of fundamentalists in the 1920s.

Acknowledgments

This book would not have been possible without generous assistance from a number of sources. The National Academy of Education and Spencer Foundation provided a postdoctoral fellowship that helped me expand, revise, and polish the manuscript. Travel grants from Binghamton University's Research Foundation and School of Education, from the University of Wisconsin's Graduate School and History Department, and from Marquette University High School in Milwaukee allowed me to include archival material from around the country. Archivists such as Lolana Thompson at Dallas Theological Seminary, Patrick Robbins and Jennifer Sackett at Bob Jones University, Joe Cataio at the Moody Bible Institute, Grace Mullen at Westminster Theological Seminary, and Wayne Weber and Bob Shuster at the Billy Graham Center Archive all extended themselves graciously to help me locate materials. The interlibrary loan staff at Binghamton University's Bartle Library helped immensely with their ability and willingness to track down obscure state law journals.

I was lucky to have not one, but two exemplary mentors throughout the years I have worked on this book. The work was guided from the outset by Ronald Numbers. Ron's early interest and continuing support and advice have been invaluable. And Bill Reese inspired both the idea for this book and its development. He has gone far beyond his original role as faculty adviser to help shape my ideas and approach.

I also need to thank colleagues who read all or part of the manuscript in various stages and offered helpful comments, ideas, and suggestions. Bill Trollinger, Jim Carpenter, Andy Cavagnetto, David Bernstein, and Eric Pullin all helped improve the work. Carol Mikoda spent many hours helping with the manuscript. Of course, the final responsibility for any errors is mine.

Ain and Virginia Laats provided unflagging support, without which I could never have devoted the required time and energy to this project. Most of all, I thank Sandra for encouraging me to write this book when it seemed a remote possibility. Even with the endless hours of talk about fundamentalists, education, and the 1920s, her energy and enthusiasm never let up.

Introduction

Tennesseans were used to hot days in July. But the temperatures in the crowded Rhea County courthouse had become so oppressive, and the crush of spectators so dangerous, that on Monday morning, July 20, 1925, Judge John T. Raulston ordered the Scopes "monkey" trial proceedings to be continued outside. Workers had set up a temporary platform under the shade of some cottonwood trees, and three thousand curious onlookers craned their necks to get a view of the trial's newest development. In this most dramatic moment of the trial, prosecutor William Jennings Bryan took the witness stand himself. What happened that day in Dayton was unexpected. It was near the end of the trial, and many of the visiting journalists had already gone home. Those newspaper writers back in Chicago, New York, and Baltimore missed the most climactic confrontation of the entire dramatic trial.[1]

When the defense called William Jennings Bryan to the stand, some members of the prosecution objected. Prosecutors did not want Bryan to be a whipping boy for fundamentalist religion. But Bryan insisted on being allowed to speak in defense of the Bible. Defense lawyer Clarence Darrow, the embodiment of the modern skeptic, relished his chance to interrogate the man many Americans revered as the peerless defender of the faith.

During the interrogation, Bryan articulated an idea close to the hearts of many Protestant fundamentalists of the 1920s. When fellow prosecutor Tom Stewart challenged the purpose of Darrow's hostile questions, Bryan seized his chance. "The purpose is to cast ridicule on everybody who believes in the Bible," Bryan thundered, "and I am perfectly willing that the world shall know that these gentlemen have no other purpose than ridiculing every Christian who believes in the Bible."

Darrow countered, "We have the purpose of preventing bigots and ignoramuses from controlling the education of the United States and you know it, and that is all."

Bryan refused to be put on the defensive. "I am simply trying to protect the word of God," he asserted, "against the greatest atheist or agnostic in the United States! [Prolonged applause.] I want the papers to know I am not afraid to get on the stand in front of him and let him do his worst! I want the world to know! [Prolonged applause.]"

The popularity of Bryan's outburst frustrated Darrow to no end. Darrow felt both pity and contempt for the poor fools who knew no better than to support such closed-minded emotional appeals. "I wish I could get a picture of these clappers," he sighed after Bryan's rousing attack.[2]

This dramatic confrontation has been worked and reworked by journalists, playwrights, and filmmakers into one of the most pivotal showdowns of American history. Unfortunately, much of the history in such plays and films as *Inherit the Wind* offered more myth and melodrama than historical fact. Nevertheless, Bryan's and Darrow's exchange of bitter comments captured the essence of the 1920s cultural conflict over the role of religion in public life.[3]

Articulating the views of many fundamentalists, Bryan argued that the ultimate moral value was fidelity to traditional interpretations of the King James Bible. Such values, Bryan thought, must remain a central part of education. In addition, Bryan protested against the growing tendency to view such traditional Protestantism with nothing but "ridicule." Bryan and many of his allies smarted at the contempt with which their beliefs were treated by many of their liberal contemporaries. Accustomed to viewing traditional evangelical Protestant faith as the revered moral compass of American education, Bryan was dismayed to see fundamentalist belief attacked as unscientific and anachronistic by America's intellectual elite.

For his part, Darrow's brief comments offered a revealing illustration of his skeptical worldview. For many opponents of fundamentalist school campaigns, such goals as the abolition of evolutionary theory could only be supported by "bigots and ignoramuses." What Darrow assumed to be the unlettered, poorly educated audience members in the Tennessee courtroom represented the kind of rural ignorance that Darrow imagined to be at the heart of fundamentalist belief. The attitudes of both Bryan and Darrow typified the two sides in cultural controversies over schooling that raged throughout the twentieth century.

It is an exaggeration to say that the Scopes trial itself gave birth to these durable controversies. But the trial did serve as the most notable incident in a decade-long fight that established new positions for what became known later in the twentieth century as America's "culture wars." Sociologist James Davison Hunter has defined those later culture wars as the struggle between "*the impulse toward orthodoxy* and *the impulse toward progressivism.*" The struggles over the nature of schooling in the 1920s described in the following pages served as the first battles in those durable cultural conflicts. The school campaigns of Bryan and his allies sketched out—if only tentatively and intuitively—the meanings of fundamentalism and its impulse toward orthodoxy.[4]

This does not mean that Americans had not experienced cultural conflict over schools before the 1920s. Catholics and Protestants had long fought, sometimes violently, over the nature of public schooling.[5] Nor does it mean that all cultural conflicts have centered solely on education. Indeed, the culture wars of the late twentieth century have often been exaggerated by scholars and journalists. But even if the "myth" of the culture wars has been overemphasized, significant cultural trenches, many of which were dug during the school controversies of the 1920s, have divided Americans throughout the twentieth century.[6]

The attention lavished on the Scopes "monkey" trial has led to some unfortunate misunderstandings about the nature of these controversies. Foremost among them has been a myopic fixation on the issue of evolution in schools. Many observers and scholars assumed that the antievolution movement represented the sum total of fundamentalist school activism during the 1920s. As this book will argue, fundamentalists carried on several other energetic campaigns to exert control over American education. In short, fundamentalists and their conservative allies sought to turn their implicit cultural assumptions about public schooling in America into explicit, legally binding educational policy. They hoped to guarantee that evangelical Protestant faith would continue to have a preferred place in America's public schools and that no school would challenge students' evangelical faith.

Some of these efforts, such as the campaign to mandate the reading of the King James Bible in public schools, had much more impressive results than the antievolution movement. Between 1913 and 1930, eleven states passed mandatory Bible-reading laws for their public schools.[7] Several more states considered similar legislation. Even in states that did not pass such laws, fundamentalists and their allies exerted enough political pressure to make elected officials squirm. Governor A. V. Donahey of Ohio, for instance, elaborately justified his veto of a mandatory Bible-reading law in 1925. Recognizing public support for this law, he assured voters that he favored Bible reading in school. However, in his opinion, the issue seemed better left in the hands of "local communities," which he encouraged to "require the reading of the Holy Bible in the schools."[8] Many local governmental bodies concurred, passing Bible-reading requirements even when states failed to do so. New York, Baltimore, and Washington, DC, all passed mandatory-reading laws during the 1920s. One contemporary activist estimated that approximately half of American cities of 100,000 or more, and an even larger proportion of smaller towns and cities, passed such laws.[9]

In addition to fighting for Bibles in schools, fundamentalists worked throughout the decade to open new schools that would reflect their side of these cultural controversies. Some of these new schools have had an enormous impact on American culture throughout the rest of the twentieth century. Bob Jones University, founded in 1926, has acted as an institutional center for a powerful network of K-12 schools nationwide.[10] The Dallas Theological Seminary, founded in 1924, has become what one historian called the "academic and theological 'Vatican'" of the movement. It became America's "most important training ground for dispensational teachers and pastors."[11] In addition to opening these hugely influential new schools, fundamentalists in the 1920s gained the allegiance of some existing institutions of higher education. The Moody Bible Institute of Chicago and nearby Wheaton College both became institutional homes for the movement. Along with a network of smaller schools, these four institutional giants helped a thriving network of new fundamentalist schools spring up during and after the 1920s, including seminaries, Bible institutes, colleges, and K-12 schools.

To be sure, fundamentalists and their conservative allies also worked to prohibit the teaching of evolution. Between 1922 and 1929, fundamentalists promoted at least fifty-three antievolution bills or resolutions in twenty-one state legislatures, plus two proposals for federal laws for the District of Columbia.[12] Five of them, in Oklahoma, Florida, Tennessee, Mississippi, and Arkansas, became laws or resolutions, and both Mississippi and Arkansas passed their laws after the Scopes trial. In fact, the most turbulent year among state legislatures was not the summer of 1925, when Clarence Darrow and William Jennings Bryan dueled to the death in sweltering Dayton, Tennessee, but 1927, when fourteen states considered similar legislation.[13] The Scopes "monkey" trial in the summer of 1925 was simply the most publicized battle in a long war.[14]

Even when fundamentalists failed to pass local or state laws to ban the teaching of evolution, they often succeeded in doctoring school districts' textbooks. Several state and local officials acceded to fundamentalist pressure to purchase only textbooks from which controversial evolutionary theory had been excised. Textbook publishers also shied away from criticism by deleting evolution from their science publications.[15]

But even the antievolution struggles of fundamentalists in this formative decade went beyond the simple issue of teaching evolutionary theory. For most fundamentalists in the 1920s, evolution symbolized a much broader web of ideas that threatened students' faith. In spite of the fact that contemporary activists and later historians have agreed that the contest focused on the teaching of evolution, many of the antievolution bills and proposals had much wider goals. For instance, Kentucky's House Bill 191 (1922) would have actually banned not just the teaching of evolution but also "Darwinism, Atheism, Agnosticism, or the Theory of Evolution." As did many of the fundamentalist school bills, this language preserved a special role for traditional Protestantism in public schools. Not only was evolution banned, but also any teaching that might shake the evangelical Protestant faith of public school children. This sweeping educational bill failed passage by only one vote.[16]

Other so-called antievolution educational proposals made similarly wide-ranging demands. In the early years of fundamentalist activist William Bell Riley's campaign to save the schools in his adopted home state Minnesota, he formed a group dedicated to banning not just evolution, but all "anti-Christian theories" from Minnesota's public schools.[17] North Carolina fundamentalists secured several county ordinances that did much more than ban evolution. The Mecklenburg County Board of Education, for instance, banned evolution and "anything that brings into question . . . the inspiration of the Bible."[18] The federal government also passed legislation widely interpreted at the time to be antievolution, but with potential for a much more transformative cultural impact. In 1924, Representative John W. Summers of Washington successfully inserted an amendment banning "disrespect of the Holy Bible" among Washington, DC, teachers.[19] In a similar vein, the antievolution Poole Bill in North Carolina (1927) actually would have banned any teaching that would "contradict the fundamental truth of the Holy Bible."[20] A proposed bill in West Virginia cut an even broader swath. That bill would have banned the teaching of "any nefarious matter in our public schools."[21] In Florida, a 1927 bill hoped to prohibit teaching and textbooks that promoted "any theory that denies the existence of God, that denies the divine creation of man, or that teaches atheism or infidelity, or that contains vulgar, obscene, or indecent matter."[22]

What result might these bills have caused? What could constitute "nefarious matter" or "infidelity"? Who decided what material was "vulgar, obscene, or indecent"? What might a teacher say that could be construed as "disrespect of the Holy Bible"? In the words of one outraged critic, these bills meant nothing less than a "far-reaching . . . revolution" in American education. Fundamentalists and their allies had set their sights on much more than banning the teaching of evolution from America's public schools. Their educational campaigns in the 1920s sought to spell out their prerogative to control the theological and cultural presuppositions of American education. Even when they did not succeed in their legislative drive, fundamentalist activism established new lines of cultural struggle and demonstrated the strength and dedication of their activist movement.[23]

In spite of many excellent academic studies of these early fundamentalists, there has been no adequate exploration of this wider educational campaign. Most historians have assumed that the controversy centered on the teaching of evolution. Instead of looking at the roots of today's conservative evangelical Protestantism, or at the origins of the controversies over the teaching of evolution, this book asks different questions: How and why did fundamentalists mobilize in the 1920s to preserve traditional Protestant education, and what effects did their campaigns have on both fundamentalism and American education?[24]

Perhaps the most significant impact of the fundamentalist educational campaigns of the 1920s was that they tore the movement apart. Fundamentalists' well-publicized educational campaigns squeezed fundamentalists into popular stereotypes about antievolution sentiment. What began in 1920 as a relatively wide alliance of conservative evangelical Protestants endured inexorable pressure to conform to a more narrowly restricted identity as a group of ignorant, rural, Southern reactionaries. Antifundamentalists used the new stereotype as a whip to chase fundamentalism out of the American mainstream. Some fundamentalists embraced the new public image. They used it as a rallying cry to attract new supporters and to motivate existing ones. Other fundamentalists, however, tried to maintain their former understanding of fundamentalism as a wider evangelical coalition, one that embraced science, demanded intellectual respectability, and included northern urban activism. By 1930, however, finding it impossible to maintain a fundamentalist movement wider than the Scopes-trial stereotype, many of these early activists drifted away from fundamentalism. They often continued their educational campaigns, especially those that had not been associated in the public mind with fundamentalism, but they usually refrained from identifying themselves or their campaigns as "fundamentalist." By 1930, as these fundamentalists abandoned the fundamentalist label, it became clear that the publicity surrounding fundamentalist school activism had irreparably divided the new movement.

Just as important, the fundamentalist school campaigns led to a transformation in American education itself. As historian Jonathan Zimmerman has noted, enduring twentieth-century conflicts over the proper role of religion in American schooling often had their roots in the 1920s.[25] At the beginning of the decade, many conservative Protestants mobilized under the banner of fundamentalism, dismayed at what they perceived to be the rapidly loosening grip of conservative Protestantism on American education. At the start of the 1920s, many fundamentalists assumed that their theology and cultural outlook formed—or ought to form—the underlying intellectual beliefs of American public education. Their school campaigns often sought simply to transform such assumptions into explicit educational law. By the end of the decade, few fundamentalists had such confidence in American education as a whole. Instead, in their eyes, public education had become a contested cultural field. In many cases, fundamentalist theology still reigned supreme. In other cases, such as at the leading public universities, fundamentalists and their conservative allies had to recognize their disappointing lack of cultural clout. The school controversies of the 1920s forced both fundamentalists and their liberal enemies to recognize each other's strength and durability. By the end of the decade, both sides realized to their surprise that they could not glibly assume they represented the overwhelming majority of Americans.

Throughout these controversies, fundamentalists confronted a wide range of opponents. Liberal Protestants and Protestant theological modernists allied with non-Protestants, advocates of academic freedom, secularists, and mainstream scientists to oppose fundamentalist school activism. Just as fundamentalists wrestled with the proper labels for their movement, so they used an array of terms to describe their enemies. Fundamentalists most commonly called their opponents "liberals," "modernists," or, less politely, "infidels," "atheists," or "skeptics." In this book, I have avoided the more derogatory terms. But I use a variety of terms to describe the antifundamentalist positions, including "liberal," "modernist," "secularist," and, more broadly, "antifundamentalist," depending on the circumstance. This should not be taken to imply that all antifundamentalists were cut from the same cloth, or that these terms are exact equivalents. As this book will describe, fundamentalists had an array of goals and attributed a variety of meanings to the fundamentalist movement. There were just as many significant differences among their opponents.

As did Clarence Darrow, many of those antifundamentalists saw their opponents as dangerous "bigots and ignoramuses." While such characterizations were often inaccurate, they became powerful weapons to discredit fundamentalism. Moreover, the stereotypes generated by the school battles of the 1920s established positions that endured throughout the culture wars of the later twentieth century. Darrow and his successors looked with scorn and bewilderment at their opponents' dogged opposition to the teaching of evolution. Bryan and the generations of fundamentalists who came after him hoped to maintain the traditional prerogatives of Protestant doctrine in America's schools. They hoped to keep a system of education that would reinforce evangelical belief. This book hopes to illuminate the meaning and development of those stubborn conflicts.

PART I

Fundamentalism and Fundamentalists

CHAPTER 1

A New Kind of Protestant

To many contemporaries, the Protestant fundamentalist movement of the 1920s seemed to come out of nowhere. In 1923, Congregational Minister Arthur B. Patten blasted the "horrific . . . dismal and devastating . . . stygian and destructive . . . desperate . . . barbarous . . . cult" of fundamentalism as an evil that had only recently "come into vogue."[1] Journalist and critic H. L. Mencken traced the roots of fundamentalism back to the passage of Prohibition in 1918. From Mencken's alarmed view in 1926, in a few short years fundamentalists had undertaken a "rapid descent into mere barbaric devil-chasing," plunging rural America "into an abyss of malignant imbecility."[2] Although the fundamentalist movement had only recently come to the attention of these critics, fundamentalism had much deeper roots. The most influential cause of the movement was the nineteenth-century intellectual revolution that included such ideas as materialistic evolution, higher criticism, and theological modernism. All three of these ideas transformed mainstream American thinking and fundamentalism emerged in part as a response to this trend.

Just as contemporaries had a difficult time understanding the history of the fundamentalist movement, so historians have disagreed on proper definitions. Since the early 1930s, popular and academic historians have described fundamentalism in egregiously misleading ways. Although the movement claimed a number of institutional bases and various types of adherents, early historians uncritically accepted hostile contemporary stereotypes of fundamentalists as rural, anti-intellectual demagogues. Early historians called fundamentalist leaders "disturbed men"[3] from a limited, "static" social environment[4] who suffered from "ignorance, even illiteracy."[5]

More recently, historians Ernest Sandeen and George Marsden have overturned this oversimplified view of 1920s-era fundamentalism. Both have conclusively demonstrated that early fundamentalists as a whole were no more rural, isolated, and uneducated than their liberal foes. However, Sandeen and Marsden each asserted a different definition of the term. Sandeen argued that fundamentalism was a modern efflorescence of the theology of premillennialism, according to which Jesus Christ would return to save a sinful world and usher in a thousand-year reign of peace and harmony. Marsden objected to such a restrictive definition. He conceded that premillennialism accounted for one important root of fundamentalism but argued that fundamentalism itself meant a wider "militantly antimodernist Protestant evangelicalism."[6]

This revision in the historical understanding of fundamentalism has generated a spate of scholarly interest in the early years of the movement. Recent historians have demonstrated that fundamentalism meant different things to different people in the 1920s. But for all of the activists involved, the causes stretched back to the cultural and intellectual revolutions of the nineteenth century. In order to understand the intensity with which both sides fought the school controversies of the 1920s, we need to review briefly those nineteenth-century transformations.[7]

For instance, Darwin's theory of natural selection had a revolutionary effect beyond the realm of the natural sciences. As soon as Darwin introduced his transmutation hypothesis with his *Origin of Species* (1859), leading American naturalists quarreled about its assumptions and implications. Within fifteen years, however, the overwhelming majority of American naturalists had accepted the premise that species had evolved. Huge segments of the American public remained unconvinced. This division between leading scientific opinion and popular thought fueled much of the vitriolic debate of the 1920s.[8]

Darwin's bombshell operated with a longer fuse among American theologians. Although many prominent voices quickly concluded that Darwin's transmutation hypothesis merely demonstrated God's method, many others restrained themselves at first from fully pondering the religious implications of Darwin's theory. Most conservative clerics assured themselves that the theory would soon be disproved by natural scientists. By 1900, however, as the tide of scientific thinking embraced the idea of organic evolution, Protestant intellectuals in America had split into two contending camps. Some favored adapting religious belief to new scientific truths and others insisted on the supremacy of revealed religion over science.[9]

The more pressing issue for many Protestant thinkers in the years immediately following Darwin's publication of *Origin of Species* was the issue of higher biblical criticism.[10] Such criticism had been hotly debated in Great Britain and America since at least 1846, when an English translation of D. F. Strauss's *Life of*

Jesus appeared. Instead of presuming the traditional beliefs of Christianity, Strauss examined Jesus as a historical person, not as the incarnate God. Following Strauss's work, other higher critics questioned the veracity and authorship of sacred texts. They pointed to glaring inconsistencies in scripture, which they attributed to fallible, human authorship. For example, critics pointed to the discrepancy between the first two books of Genesis. In the first chapter, God created Adam after the beasts, while in the second, He created Adam before them. Using this kind of evidence, higher critics argued that the book was cobbled together by talented, but still merely human, editors. This approach to scripture radically challenged many fundamental assumptions of traditional American Protestantism. By the 1920s, popular and intellectual outrage against this assault on the supernatural status of the Bible formed one important source of fundamentalist support.[11]

A related nineteenth-century idea that fueled the twentieth-century fundamentalist movement was theological modernism. Prominent American theologians had articulated this idea since at least the 1870s. In brief, it entailed "the conscious, intended adaptation of religious ideas to modern culture . . . the idea that God is immanent in human cultural development and revealed through it . . . [and] a belief that human society is moving toward realization . . . of the Kingdom of God."[12] Just as Darwin's transmutation hypothesis did, this approach to Protestantism split the Protestant community. For many Americans, theological modernism allowed them to maintain their faith in an age of rapid intellectual and social turmoil. For fundamentalists and other conservative Protestants, however, theological modernism was merely an abstruse new articulation of ideas as old as Moses. For many of those who became fundamentalists in the 1920s, the call to adapt Christian belief to modern American society was no different than the temptation of the Jewish people in the Old Testament to adopt local gods and religious practices, and to stray from the strict worship of the God of Abraham.[13]

The effects of these intellectual revolutions combined to threaten the core beliefs of many Americans. Throughout the mid- and late-nineteenth century, evolutionists argued that religion must be adapted to accept Darwin's radical theories about the origins of species. Higher critics dissected the Bible in a quest to prove that it was nothing more than an edifying collection of Semitic myths. Theological modernists urged Protestants to be reasonable about their religion, to accept major changes to stay in step with developments in the wider culture. In each case, modern American culture asked Protestants to question some of their bedrock beliefs.

By the 1920s, such modern, liberal views had come to represent the mainstream in American thought. At the Divinity School of the University of Chicago, for instance, theological modernism, higher criticism, and theological evolution had become the norm. Two leaders of the school, Dean Shailer Mathews and Professor Shirley Jackson Case, became prominent advocates of the new theology. In 1919, in *The Revelation of John*, Case analyzed the Book of Revelation as a political allegory about the declining Roman Empire, with meanings only relevant to the time the book was written. In 1924, Mathews published *The Faith of Modernism*, an impassioned defense of theological modernism.[14]

The widespread acceptance and promotion of these ideas on college and university campuses occurred at a time of rapid school expansion. College attendance numbers exploded between 1870 and 1930.[15] Student enrollment also increased in leaps and bounds at elementary and secondary schools. In some public school districts, these increases came as a shock during the boom years of the 1920s. The school board of Atlanta, Georgia, for instance, was caught unprepared in 1925 when over two thousand more students than expected showed up for the first day of school. The 45,466 students in Atlanta's public schools in 1925 filled the schools to bursting. Only six years earlier, the local schools had accommodated less than half that number.[16] Atlanta was not an exception. All over the country, more Americans were spending more time in school during the 1920s. Evangelical students and their parents became concerned, understandably, with the curricula taught at those schools, and the effect it might have on their careers, their social standing, and, for many, their immortal souls.[17]

In large part, the fundamentalist movement of the 1920s emerged as a popular reaction to the neo-Darwinism and theological liberalism that followers believed had taken root in many denominations and school systems. However, it would be too simple to portray early twentieth-century fundamentalism as merely a knee-jerk reactionary movement intent on recreating an imagined Protestant golden age in America. A fuller understanding of fundamentalism must also include other important roots. As historian George Marsden has demonstrated, fundamentalism emerged out of a combination of elements, including the theology of dispensational premillennialism, the holiness revival movement, and a mix of ideas about the proper relationship of Christianity to culture.[18]

Many fundamentalists of the 1920s viewed the world through the lens of premillennial dispensationalist theology. Briefly, this theology developed from the widespread Protestant belief in the Bible as an inerrant text, as argued most forcefully by Presbyterian theologians at the Princeton Theological Seminary.[19] British evangelist John Nelson Darby brought dispensational premillennialism to the United States in the years following the American Civil War as the theology of his Plymouth Brethren sect. Darby found his American audiences largely uninterested in joining the Plymouth Brethren, but his theology attracted many important converts. American evangelical author William E. Blackstone popularized many of the tenets with his influential *Jesus is Coming* (1908). More important, the commentary of Cyrus Scofield in his widely read study Bible enshrined the beliefs of dispensational premillennialism in the minds of many American Protestants as part of Holy Writ itself.[20]

Darby and a small group of dedicated American theologians, mainly Presbyterians and Northern Baptists, began to coalesce their theological movement in 1876 by founding a series of prophecy conferences that came to be known as the Niagara Conferences.[21] They started with the belief that the Bible was entirely inerrant. Seeming contradictions, such as the clash between Jesus's and Paul's teachings and the fact that many Old Testament prophecies did not refer to a Christian church were explained by dividing history into a series of dispensations. The current period, the "Church Age," was a time of declining morality, a time when nominal Christians were winnowed away from the pure remnant. This division of sacred history into dispensations cleared up the apparent textual contradictions. The earlier age of the law still held during Jesus's lifetime on Earth, so his teachings reflected an appropriate justification by law. Paul's thinking, in contrast, reflected the new rules of the Church Age. The ignorance of Old Testament prophets about the Christian church resulted merely from the fact that they were understandably unaware of the mystery of the current Church Age.

In addition, some believers in this theology held that endtime prophecies, such as those in the Books of Daniel and Revelation, referred literally to future events. For instance, the phrase "Christ's return" should not be interpreted as a metaphor, or as a symbol of spiritual conflict, but as a more mundane description of the future: Christ will return physically to combat the forces of evil and sin. He will begin by calling the few true believers to join him. This "secret rapture" was one of the novelties of Darby's theology. After the true Christians met Christ "in the air," seven years of tribulation would follow, all of which were precisely described in the Books of Daniel and Revelation. At the end of this time, Christ will conquer then usher in the millennium with a thousand years of peace and human harmony.[22]

Dispensational premillennial theology tended to ignore social and political issues. Since this world was a wrecked vessel, waiting only for Christ's return, it made sense for premillennialists to separate themselves from a corrupt society. This theological tendency has led some historians to the paradoxical conclusion that real fundamentalists never engaged in the kind of cultural and political conflict that fundamentalism became known for.[23] In fact, however, early fundamentalist theology was a complex creation that often encouraged social activism. Many of the same prophecy students who were involved in the Niagara Conferences also participated in activist traditions, especially the nineteenth-century Holiness movement. This movement stressed the idea that the Holy Spirit was being poured out in the present. The idea had deep roots, stretching back at least to John Wesley, one of the founders of Methodism. In the nineteenth century, evangelists such as Phoebe Palmer popularized the notion of living a "victorious life." Holiness advocates spoke of being filled with "power for service," and of ecstatic experiences of the Holy Spirit in their lives. This movement inspired many 1920s-era fundamentalists with "a profound personal experience of consecration, a filling with Spiritual power, and a dedication to arduous Christian service."[24]

The rise of theological modernism, higher criticism, and theological evolutionism brought these two movements together. Many "denominational traditionalists," especially among established denominations such as the Northern Baptists and Northern Presbyterians, feared the newfound influence of these ideas in denominational seminaries. In an effort to defend their traditional Protestant beliefs, many of these conservatives eagerly allied with evangelists of other denominations to form fundamentalism as a recognizable social and theological movement.[25]

Another important root of the vigorous movement culture of 1920s-era fundamentalism was the fertile mixture of activists of different backgrounds. Fundamentalists held differing beliefs about the proper role of religion and culture. Some of the leading fundamentalist voices of the 1920s, such as prophecy writer and editor Arno Gaebelein, advocated a strict separation from a corrupt and corrupting American culture. Other activists, such as Moody Bible Institute leader James M. Gray, focused primarily on the need for revival, but believed that cultural reform could be a beneficent result of widespread soul saving. Still others combined traditional American Protestant themes with an emphasis on social reform. William Jennings Bryan exemplified this tendency; he clung vigorously to the idea that American civilization depended on the bedrock morality of Protestantism. J. Gresham Machen, a leading Presbyterian theologian at Princeton Theological School during most of the 1920s, voiced an even stronger role for Christianity in American culture. Machen argued that American culture must be transformed by Christianity; there must be a "consecration of culture."[26]

As historian George Marsden convincingly argued in *Fundamentalism and American Culture*, these four roots—the theology of premillennial dispensationalism, the Holiness revival, the perception of besieged traditional beliefs, and the blending of several views of the proper relationship between Christianity and American culture—combined in the 1920s to create a disparate theological movement to save American Protestantism from itself. Those four roots describe well the fundamentalist movement that fought for control of major Protestant denominations. However, the fundamentalist movement also fought a broader battle in the public sphere to save all of America from itself. As the next chapter will describe, activists and observers in the 1920s did not limit their definitions of fundamentalism to militantly antimodernist evangelical Protestants. In practice, as often happens with political and cultural labels, "fundamentalism" took on a spectrum of meanings fitted to contemporary struggles.[27]

Although the ideas that made up the movement went back at least a generation, conservative evangelical Protestants did not consider themselves fundamentalists until the 1920s. The name itself, along with a sense of common identity and purpose, was a novelty of the controversy-filled 1920s. Use of the root "fundamental" among conservatives went back at least to the publishing of *The Fundamentals*, a twelve-volume set of booklets that the Stewart brothers, two wealthy California oilmen, published between 1910 and 1915. Although the editors took a relatively mild and ecumenical tone by the standards of the acrimonious controversy that developed in the 1920s, the goal remained similar to the later fundamentalist position: to promote certain beliefs as the basic truths of Christian belief, including the inerrancy of the Bible, the deity of Christ, the virgin birth of Christ, His atonement for sin, and His bodily resurrection. By distributing approximately three million volumes free of charge to Protestant pastors, Young Men's Christian Association (YMCA) leaders, and congregations, the Stewarts hoped to push back the growing influence of theological modernism.[28]

The term fundamentalism itself was not coined until July 1, 1920, when Curtis Lee Laws, the editor of a Baptist newspaper, the *Watchman-Examiner*, first used it to describe the growing protest against modernism among conservative evangelicals. As Laws later recalled, he was worried about some of the negative terms being used by liberals, such as "'literalists,' 'dogmatists,' 'separatists,' 'medievalists,' 'cranks,' 'ignoramuses,' and 'ku-kluxes.'" He hoped that the catchy term fundamentalism would supplant all those derogatory terms and unite all the Protestants who believed in the five fundamental beliefs of the 1876 Niagara Conference.[29]

In 1922, the fight between fundamentalists and modernists absorbed much of the attention of Northern Presbyterians and Northern Baptists. The spark that ignited the hottest part of the controversy came on May 21, 1922. The parishioners and guests at New York's First Presbyterian Church heard a fiery sermon that morning, delivered by one of the leading voices of the modernist, liberal wing of the denomination. Ironically, the popular pastor Harry Emerson Fosdick was not even a Presbyterian, but a Baptist. In his aggressive sermon, "Shall the Fundamentalists Win?" Fosdick threw down the gauntlet to fundamentalists of any denomination, lambasting them as "essentially illiberal and intolerant."[30]

Presbyterian conservatives and fundamentalists fought back. Philadelphia pastor Clarence E. Macartney quickly replied with a rebuttal sermon: "Shall Unbelief Win?" At the General Assembly of 1923 in Indianapolis, the New York Presbytery was slapped on the wrist for supporting Fosdick, and the Assembly affirmed a five-point 1910 doctrine of Presbyterian faith: an inerrant Bible, the virgin birth of Christ, His substitutionary atonement, His bodily resurrection, and the authenticity of miracles.[31]

By the next year, however, Presbyterian liberals and moderates regrouped and took control of the denomination. At the General Assembly of 1924 liberals presented the "Auburn Affirmation" with 1,300 signatures. This document denounced the previous year's five points as un-Presbyterian, due to the fact that they exceeded the requirements of orthodoxy set down in the traditional Westminster Confession of Faith. A report from the Permanent Judicial Commission forced Fosdick to resign his pulpit, but it did not require that all Presbyterian pastors adhere to the five-point doctrine. Fundamentalist attempts to force a review of that ruling were defeated by a wide margin.[32]

Northern Baptists experienced a similar struggle. In 1922, William Bell Riley, founder of the World's Christian Fundamentals Association and a leading voice for fundamentalism, pushed unsuccessfully for a more conservative creed at the 1922 national Convention. Unfortunately for Riley and other Baptist fundamentalists, liberal pastor Cornelius Woelfkin offered an irresistible counterproposal. Woelfkin won the day when he proposed that the New Testament was the only creed Northern Baptists needed. Frustrated Northern Baptist fundamentalists founded a rump group, the Baptist Bible Union, but most remained members of the denomination.[33]

Organized and energized by their fight for control of their denominations, many fundamentalists put renewed energy into their campaign for control of America's schools.[34] Schools, after all, had been one of the first areas of concern for many fundamentalists. At the founding meeting of the World's Christian Fundamentals Association in 1919, for instance, leader William Bell Riley warned his audience about schools that "use text books or employ teachers that undermine the faith in the Bible as the Word of God and in Jesus Christ as God manifest in the flesh."[35] Other voices at that same convention pinpointed a more specific threat in the schools, namely, the theory of evolution. Charles A. Blanchard, president of Wheaton College in Illinois, denounced such teaching as "not only unsupported by any unquestioned facts and therefore totally unscientific, but . . . a distinct denial of the Bible account of the creation of man, the beginning of sin, the plan of salvation and the extension and triumph of the Christian religion in the world."[36]

William Jennings Bryan, the most prominent fundamentalist leader of the 1920s, agreed. By 1921, Bryan had come out of semiretirement to battle against the teaching of evolution as a fact in schools. He began traveling the country delivering two polished stump speeches: *The Menace of Darwinism,* and *The Bible and Its Enemies.* In each of these speeches, Bryan used his practiced popular style to condemn the teaching of evolution.[37] In *The Bible and Its Enemies,* he painted a grim picture for his audiences. He described "a boy reared in a Christian home, learning the first child's prayer and then the Lord's Prayer; he talks to God, asks for daily bread, pleads for forgiveness of sins, and desires to be delivered from evil. . . . Then he goes off to college and a professor takes a book six hundred pages thick and tries to convince him that his body is a brute's body. 'See that point in the ear? That comes from the ape,' etc." These concerns about American schools soon led to the most prominent public controversies of the fundamentalist movement in the 1920s.[38]

But fundamentalists fought for more than just control of denominations and schools. Many prominent fundamentalists led personal crusades against public vice. One of the best known of these crusaders was New York pastor and activist John Roach Straton. Straton seized the public eye in New York City with his loud, colorful campaigns against dancing, boxing, and Sabbath-breaking. Straton gained national fame when he agreed to supply fundamentalist commentary to the 1921 Jack Dempsey-Georges Carpentier prizefight, a spectacle Straton denounced as a "moral carbuncle."[39] Straton also fought to see "red-light districts . . . eliminated, prostitution minimized, and venereal diseases conquered."[40]

The fundamentalist movement in the 1920s also claimed the allegiance of the greatest itinerant evangelist of the Great War years, Billy Sunday. Sunday fiercely opposed the teaching of evolution in schools and he attacked theological modernists with unmatched venom. He traveled to Memphis as the Tennessee state legislature considered its antievolution Butler Bill. In a speech to more than five thousand, he praised the legislative work of the antievolutionists, "for its action against that God forsaken gang of evolutionary cutthroats."[41] However, by the time of the Scopes trial, Sunday's ability to attract audiences had declined significantly from its peak in 1918. He remained interested in educational issues until the end of his life in 1935, even joining the board of trustees of Bob Jones College in the 1930s. However, he did not participate in prominent fundamentalist organizations. Rather, he continued his diminished touring, delivering what had become almost perfunctory sermons to smaller and smaller crowds in smaller and smaller towns about the dangers of alcohol, immigration, and political radicalism.[42]

A few leading fundamentalists, such as Seattle's Mark Matthews, paid little attention to educational campaigns. Matthews occasionally warned his radio audience about the dangers of secular education, saying, "We will never permit our educational institutions to make an attack on God, upon the Bible, upon the faith of our children." But during the Scopes trial, he shied away from public support of his fellow Presbyterian William Jennings Bryan.[43] He made few public statements about the trial, preferring to focus on his mission to create in Seattle a "righteous community," free from the modern scourges of "amusement mania or pleasure insanity."[44]

The different emphases of these leading fundamentalists demonstrate the complexity and diversity of the movement itself during the 1920s. In spite of such diversity, some common ideas united this disparate movement. One common trait was a growing sense of profound cultural disjunction. By the 1920s, both sides in the long nineteenth-century debates over evolution and theology had developed such distinct ways of understanding the world and of humanity's role within it that they often claimed to be unable to understand how anyone could possibly hold the opposing view. Some of these claims must have been sheer grandstanding. But to some extent, fundamentalists and their foes had developed such distinct cultures that the very ontological and epistemological presuppositions of the two parties had become mutually threatening. For fundamentalists, for example, humans existed as the product of a loving God. To assert, as many materialistic evolutionists did, that humanity had merely evolved by chance threatened the taproot of that belief. Similarly, fundamentalist belief, at its core, depended on the idea that knowledge of the world developed from revelation. In order to understand the world, the source of that revelation, the Bible, must remain the starting point of investigation. To assert, as many theological modernists did, that the Bible was merely an assemblage of historic documents cobbled together by fallible human editors challenged the very core of this epistemology. Similarly, the claim of organic evolution implied that the results of human inquiry could supersede biblical teaching. If fundamentalists were to accept theological modernism and organic evolution as true, then they would have had to abandon the center of their modes of being and knowing. Understandably, most were loath to do so, especially when it came to the education of their children. If the children of fundamentalists were taught the assumptions of the modernists and evolutionists, then they would necessarily be divided, in their very ways of being and understanding, from their own parents. And this possibility inspired fundamentalists by the thousands to join the educational crusades of the 1920s.

For 1920s liberals, the opposite was the case. Truth could only be trusted if it had been tested according to modern scientific methods. Across this cultural divide, each side had a difficult time understanding the arguments of the other. Both sides often shook their heads and concluded that their opponents' views were absurd. One liberal opponent of Kentucky's school antievolution bill, for instance, lamented that without evolution, there "would be little left for [schools] to teach."[45] Another described the antievolution legislation as "unwise, absurd, ridiculous."[46]

Fundamentalists shared this confusion about their opponents. One southern Presbyterian concluded that only "lunatics" could oppose fundamentalist school policy.[47] William Jennings Bryan denounced evolution as "laughable" in an early opinion piece commissioned by the *New York Times*.[48] In one of his speeches of the period, Bryan expanded on this opinion. "I do not object to an absurd hypothesis when it does not hurt anyone," Bryan charged, articulating the view of many who had been drawn into fundamentalist school controversies, "But these imaginings are not only groundless and absurd but harmful."[49] Similarly, when New York fundamentalist leader John Roach Straton reviewed Sinclair Lewis's fundamentalist-bashing novel *Elmer Gantry*, Straton called the novel "the apotheosis of the absurd."[50]

Some of these charges must have been nothing more than crude attempts to discredit the opposition. But it is clear that committed fundamentalists and liberals confronted each other over a profound cultural divide, one that made the others' views incomprehensible. This lay at the root of the bitter school controversies of the 1920s. For both fundamentalists and liberals, it had become a high-stakes game, and neither side could afford to lose.

CHAPTER 2

What's in a Name?

Although fundamentalists may have agreed that the intellectual pretensions of their opponents were preposterous, they often had a difficult time on agreeing to much more. In the first few years of the 1920s, both fundamentalists and the wider public struggled to understand the new movement. No less than later historians did, fundamentalists and their contemporaries often disagreed about what fundamentalism meant. Some leading fundamentalists attempted to assert a definition on the movement unilaterally. Other leaders avoided using the term. And many contemporaries used the term to refer to a broad assortment of conservative trends in politics, culture, and religion. As one Baptist editor complained in 1923, "Millions of people have been confused by this controversy."[1]

With good reason. To borrow a cartographic metaphor from sociologist of science Thomas Gieryn, the first fundamentalists and their enemies were engaged in the complicated process of socially constructing the boundaries of a space that was recognizably fundamentalist. This entailed a never-ending process of "boundary-work," during which all actors tried to impose a cut-and-dried definition on necessarily porous and shifting terrain. As Gieryn has written about the nature of science, contemporary understandings of the meaning and definition of the fundamentalist movement in the 1920s achieved "authority precisely from and through episodic negotiations of its flexible and contextually contingent borders and territories." Fundamentalists and their contemporaries subjected the meanings of the new movement to constant scrutiny, debate, and negotiation.[2]

Early historians of fundamentalism made the mistake of accepting and reifying one side of this negotiation over the meanings of fundamentalism as the essence of fundamentalism. They accepted the assertions of the enemies of fundamentalism as the simple truth. Thus, liberal accusations of irrationality and rural isolation dictated the historical interpretation of the movement for decades. The real historic situation was much more complicated. Fundamentalist identity, like other social identities, emerged as the product of an implicit, protracted negotiation between interested parties.[3]

These definitional negotiations were more than mere scholastic disputes. Contending definitions of fundamentalism determined a good measure of the success and failure of fundamentalist educational policies. As this chapter will describe, many leading fundamentalist activists came to different conclusions about the meaning of their movement. Several leaders attempted to assert rigid boundaries of fundamentalism. Yet in practice, many conservatives during the 1920s called themselves fundamentalists without too much thought for such niceties of definition. Thus, in spite of meticulous boundary-work on the part of many fundamentalists, many conservatives, including Pentecostals, Ku Klux Klan members, conservative Lutherans, and even conservative Catholics often used the term fundamentalist to describe themselves.

When it came to the public campaigns to change schooling, these shifting definitions played a decisive role. By the end of the 1920s, many contemporaries simply equated fundamentalism with the antievolution movement. But fundamentalist school campaigns often included a broad spectrum of enthusiastic conservative support from activists who did not consider themselves fundamentalists. Even many liberal Protestants supported the drive to mandate the reading of Bibles in public schools, as we will see in Chapter 8; and although many conservative Baptists and Presbyterians did not join the fundamentalist side of their denominational controversies, they participated eagerly in campaigns to maintain the dominance of Protestant theology in public schools. Attempts to draw bright dividing lines between fundamentalists and nonfundamentalists break down in this thicket of activism and self-identification. Nevertheless, in order to understand the many meanings of fundamentalism during the 1920s, this chapter will attempt to untangle some of these definitional threads.

Any examination of the meaning of fundamentalism in the 1920s ought to begin with William Bell Riley. His talent for organization catapulted him to leadership of the movement when he founded the World's Christian Fundamentals Association (WCFA) in 1919. Riley, the popular pastor of the First Baptist Church in Minneapolis, worked to create a role for himself as ultimate arbiter of self-definition for the fundamentalist movement in the 1920s. Riley's magazine, *Christian Fundamentals in School and Church (CFSC)*—which he renamed *The Christian Fundamentalist* in 1927 to assert its status as the leading fundamentalist magazine—became a leading organ of the movement.[4]

In the first years of the 1920s, Riley clung to other labels, such as "orthodox" and "evangelical" to describe his position.[5] But by the beginning of 1923, Riley committed himself to the fundamentalist movement and worked hard to define the proper boundaries of fundamentalism to himself and to his readers. For instance, in the January 1923 issue of his magazine, Riley ran six articles defining fundamentalism. In October of the same year, Riley published five such articles: "Fundamentalism—The Word That Has Won Its Way"; "Tampering with the Formulae; or, Why I Am a Fundamentalist"; "Fundamentalism: A Call Back to the Bible"; "A Fundamentalist Church on the Foreign Field"; and "Successful Fights for Fundamentalism."[6]

Each of these articles offered readers a definition of the controversial new movement. In one article, Riley described the results of a survey about the nature of fundamentalism. The survey had asked both leading fundamentalists and leading modernists about the meanings of the new term.[7] Predictably, fundamentalists defined the movement as a heroic attempt to defend the eternal truths of the Christian religion. Modernists attacked it, in the words of one writer, as "sterilizing . . . the doctrine of despair."[8] In other articles in this issue, Riley noted that the use of "fundamentalism" was spreading far beyond what he considered the true fundamentalist movement. For Riley, the label's correct use was only to describe his conservative evangelical Protestant protest movement, noting that everyone from advertisers to politicians used "fundamentalism" to suit their own purposes.[9] Elsewhere in the October issue, another fundamentalist author defined the new movement as "a call back to the Bible from which many have been led astray. It is an agitation in defense of the historic faith of the great evangelical denominations." As did Riley, this writer noted that the label had been claimed and used in many other ways. He hoped to assert a single proper meaning for the term.[10]

Riley soon began to promote himself as the spokesman for fundamentalism to the American public at large. He asserted that the nine-point creed of the WCFA represented the official creed of the movement. The 1919 creed is so central to understanding Riley's attempt to define fundamentalism that it is worth including *in toto*:

   I. We believe in the Scriptures of the Old and New Testaments as verbally inspired of God, and inerrant in the original writings, and that they are of supreme and final authority in faith and life.

  II. We believe in one God, eternally existing in three persons, Father, Son and Holy Spirit.

 III. We believe that Jesus Christ was begotten by the Holy Spirit, and born of the Virgin Mary, and is true God and true man.

IV. We believe that man was created in the image of God, that he sinned and thereby incurred not only physical death, but also that spiritual death which is separation from God; and that all human beings are born with a sinful nature, and, in the case of those who reach moral responsibility, become sinners in thought, word, and deed.

V. We believe that the Lord Jesus Christ died for our sins according to the Scriptures as a representative and substitutionary sacrifice; and that all that believe in Him are justified on the ground of His shed blood.

VI. We believe in the resurrection of the crucified body of our Lord, in His ascension into heaven, and in His present life there for us, as High Priest and Advocate.

VII. We believe in "that blessed hope," the personal, premillennial and imminent return of our Lord and Saviour Jesus Christ.

VIII. We believe that all who receive by faith the Lord Jesus Christ are born again of the Holy Spirit and thereby become children of God.

IX. We believe in the bodily resurrection of the just and the unjust, the everlasting blessedness of the saved, and the everlasting, conscious punishment of the lost.[11]

Riley worked throughout the early years of the 1920s to promulgate this nine-point creed as the official definition of fundamentalism. But throughout those early years of fundamentalism, he did not insist that all of his fundamentalist allies adopt this creed. He reached out to prominent leaders, such as Bryan, in spite of Bryan's well-known disavowal of the theology of premillennialism, and thus of the seventh point of Riley's creed.[12] Like other early fundamentalists, Riley wanted to build a wide evangelical coalition around the new label. Unlike some of his fundamentalist allies, however, Riley committed himself and his organization entirely to the new movement. When public attention forced a constriction of the meanings of fundamentalism, Riley was too devoted to the term to abandon it, in spite of his bitterness about new connotations of bigotry and anti-intellectualism. Instead, in the years following the Scopes trial, Riley would continue to fight for control of the meanings of fundamentalism.

J. Frank Norris, the pastor of the First Baptist Church in Fort Worth, eventually came to identify with fundamentalism as closely as Riley had. His early experience with the new movement was similar in many ways to that of Riley. Like Riley, he did not immediately seize the new label to describe himself. Once he did, however, he claimed to be the gatekeeper for the new movement. Just as

Riley had, Norris unilaterally asserted a definition of fundamentalism. The tension between these two self-appointed guardians of fundamentalism remained productive as long as fundamentalism maintained relatively wide boundaries. However, as we will see, after the Scopes trial, restricting boundaries made control of the definition of fundamentalism a more contentious issue and the two leaders could no longer work together.[13]

Norris had always associated himself with the conservative side of evangelical disputes. He believed strongly enough in the issue to become a charter member of Riley's World's Christian Fundamentals Association in 1919. Like Riley, though, Norris did not immediately use "fundamentalism" to describe his own beliefs. For Norris, the controversy at first seemed to be a purely Northern problem. As late as 1922, Norris's magazine, *The Searchlight*, made this distinction very clear: "That there are two distinct groups of Baptists within the Northern Baptist Convention is a fact too well-known to need proof. The terms 'conservative' and 'liberal,' 'fundamentalist' and 'modernist' sufficiently indicate the denominational consciousness of these two groups." Until early 1923 Norris assumed that all Southern Baptists were conservative; therefore there was no need to describe them as fundamentalists. For Norris, the label Southern Baptist included all the important tenets of fundamentalism; a Southern Baptist fundamentalist would have been a redundancy.[14]

Norris first explicitly accepted the label in 1923, and immediately began to assert a heroic identity for fundamentalism.[15] Just as Riley's had, Norris's promotion of the new movement took a central role in all of his writing and preaching. For instance, he contrasted the bravery and fortitude of a "thorough-going dyed-in-the-wool Fundamentalist" to the hypocrisy and laziness of less dedicated Southern Baptists.[16] He also began to try to define fundamentalism to the wider public. In response to a negative article in the *Dallas News*, he asked for space to present the fundamentalist side of the controversy. In his answer, Norris tried to define what fundamentalism meant. He repeated five central tenets that he had often referred to in public addresses: "First the inerrancy of the Scriptures. Second, the virgin birth. Third, the deity of Christ. Fourth, the substitutionary atonement. And, fifth, the imminent, physical and literal return of the Lord."[17]

For Norris, fundamentalism was a straightforward affair. Those who adhered to his creed were welcome to join Norris in the fundamentalist movement. Those who opposed him were dubbed opponents of fundamentalism. Even the storms of publicity surrounding the Scopes trial left Norris unruffled. The new stereotyped popular image of fundamentalism did not discomfit him much. In fact, Norris became for many the epitome of the newly stereotyped image in 1926 when he shot a rival to death in Norris's own office. Although he loudly and successfully claimed innocence, the image of a gun-toting, trigger-happy Southern preacher leading the fundamentalist movement did a great deal to cement the restricted stereotype as the true definition of fundamentalism.[18]

James M. Gray, the leader of the Moody Bible Institute in Chicago (MBI), the leading institution of fundamentalism in the 1920s, grafted distinctly different meanings onto his understanding of the movement. Gray was slower to accept the new label, and he never agreed with Riley's creed. In the end, MBI leaders such as James M. Gray accepted the label with some reservations once the differences between fundamentalism and a wider conservative evangelical Protestantism seemed to become negligible. However, once fundamentalism became restricted to the popular stereotype in the later years of the 1920s, James Gray quietly abandoned the term.

One of the reasons for this different approach to fundamentalism was the organizational difference between Gray's institutional MBI and Riley's or Norris's one-man shows. The MBI had been a leading evangelical institution since 1886, and Gray had prosaic worries such as payroll, physical plant, and a cautious board of directors to consider. The MBI also had a greater scope of activity than the WCFA. In addition to its central institutional goal of instilling in students a thorough dispensational understanding of the Bible, it hosted innumerable conferences, published a monthly magazine, and encouraged missionary work and domestic evangelical revival. In addition, the MBI was associated with its own publishing house, and, by 1926, it had even begun operation of its own radio station.[19]

Throughout the first years of fundamentalism's existence, Gray struggled to understand the meanings of the new term. Unlike Riley and Norris, he held on tenaciously to his preferred "evangelicalism" to describe the conservative side of the controversies, and it was only with regret that he accepted "fundamentalism."[20] By the end of 1922, Gray conceded that the new label had come to include the MBI's style of conservative biblical evangelicalism.[21]

For Gray, and for the wider MBI community in the early 1920s, fundamentalism was used as the appropriate name for the antimodernist revival already under way at the MBI, but not necessarily for Riley's nine-point creed. In other words, Gray accepted the label with reservations once fundamentalism had become the accepted label for the entire conservative evangelical movement. However, just as they had for Riley, Gray's attempts to build a workable self-understanding as a fundamentalist were severely shaken by the bitterness of the fundamentalist school controversies. Unlike Riley, Gray and the MBI never committed themselves so firmly to a strictly fundamentalist identity that they could not quietly move away from fundamentalism once its boundaries had become uncomfortably restricted following the Scopes trial. As we will see, in the years following the public scrutiny resulting from media circuses like the one surrounding the Scopes trial, Gray did just that.

Like the MBI, the Bible Institute of Los Angeles (now Biola University) only came around slowly to an acceptance of itself as a fundamentalist institution. Many in the Biola community initially preferred such terms as evangelical and orthodox. By 1923, however, Biola had accepted "fundamentalism" as a legitimate label for its own theology and activities. The struggle was very similar to that carried on at the MBI. Just as Gray only accepted the label once it had become successfully established, the leaders of Biola eventually accepted the new label as a fait accompli. There were many reasons for this similarity. Biola's founders recruited their first dean, Reuben A. Torrey, away from his position as dean of the MBI, and when they proposed their own magazine, *The King's Business*, they consciously imitated the *Moody Bible Institute Monthly*. Biola also proudly copied the MBI's programs of evangelism and Bible instruction.[22]

In early 1923, Biola began to identify with the fundamentalist movement. In a reply to a critical liberal article, "An Open Letter to a Fundamentalist," *The King's Business* published "An Open Letter to a Modernist." This letter challenged the liberal definition of fundamentalism, and offered a heroic definition instead:

> You call me a "Fundamentalist." I esteem this an honor. . . . The very name "Fundamentalist" proclaims faith in the Son of God, whose teachings I deny any man the right to modify.
>
> Cordially yours,
> A FUNDAMENTALIST.[23]

As did Gray of the Moody Bible Institute, the leaders of the Bible Institute of Los Angeles accepted the new label once it seemed to have become the standard term for the conservative, biblical side of Protestant controversy. Unlike Riley or Norris, most members of the Biola community did not assert simple creeds for the movement. Nor did they presume to dictate the exact meanings of the label. Rather, they understood the movement to be a welcomed revival of conservative evangelicalism. For the leaders in Los Angeles, fundamentalism meant a fight for the traditional primacy of an inerrant Bible, but not necessarily for the specific creeds of leaders such as Riley or Norris.

One of the most prominent leaders of the fundamentalist movement in the 1920s negotiated an even more tenuous identification with the movement.

William Jennings Bryan helped turn fundamentalism into a household term when he volunteered to assist the prosecution at the Scopes trial in 1925. Though many of his former political allies were confounded by his new campaign, the move made sense for Bryan. It marked the end of a thirty-year career in the national spotlight, which had begun with his promotion of the "Free Silver" campaign in his presidential candidacy of 1896. He had gained further fame by dramatically resigning from his post as secretary of state over the *Lusitania* controversy. The controversy over the teaching of evolution in the public schools roused him from semiretirement in Florida and his prominent place in the public eye guaranteed his leadership of the antievolution insurgency.[24]

Despite leading this popular campaign, Bryan carefully avoided the term fundamentalism. He never used it in his public speeches or published writings, even though both sides agreed that Bryan was the undisputed leader of the fundamentalists. Before the word "fundamentalism" caught on, liberal critics often identified the entire antimodernism movement as "Bryanism."[25] One liberal argued that "Mr. Bryan is Fundamentalism. If we can understand him we can understand Fundamentalism."[26] Admirers were no less prone to give Bryan the honor of leadership. An editorial in *The King's Business* praised Bryan as "the outstanding layman leader of the Fundamentalists of the United States."[27] *The Searchlight*, J. Frank Norris's newspaper in Fort Worth, Texas, called Bryan the "Great Fundamentalist."[28] Bryan was even offered the leadership of the Laymen's League of the World's Christian Fundamentals Association, which he politely declined.[29]

Bryan himself preferred other ways to describe the two sides of the controversy. For instance, he often counterposed "Evolutionism" to "Christianity." In speeches to fundamentalist audiences, he might refer to his delight at being able to speak to an audience of "only Christians," all of whom could be counted on to oppose theological modernism.[30] Even the speech he had planned for his closing remarks at the Scopes trial never referred to "fundamentalism" but described the conflict as one between "unbelievers" and "Christians." Because virtually everyone but Bryan was using fundamentalism by this time to describe the conservative side of these controversies, his reticence suggests some degree of conscious avoidance of the term.[31]

At times, Bryan implied that he was not a fundamentalist. He thanked Riley for "the opportunity the Fundamentalists have given me to defend the faith," suggesting a distance between himself and fundamentalism.[32] This reticence has prompted Michael Kazin, Bryan's most recent biographer, to conclude that Bryan was not a fundamentalist at all.[33] But Bryan's relationship to the term fundamentalism was more complicated than that. In some situations he acknowledged his leadership role in the fundamentalist movement. In one off-the-cuff comment to a reporter, for instance, he noted that "people often ask me why I can be a progressive in politics and a fundamentalist in religion. The answer is easy. Government is man-made and therefore imperfect. It can always be improved. But religion is not a man made affair." Bryan wanted to be seen as a fundamentalist, yet also wanted to be able to deny the label.[34]

His coyness with fundamentalism must be understood as a political strategy from a master of the game. He was willing to implicitly accept the leadership of the fundamentalist movement in order to help his antievolution campaign. But, just as he had done thirty years earlier with the Populist Party, Bryan insisted on holding himself somewhat aloof. He did not want his own influence to be limited to that of an inchoate grassroots movement. If Bryan had survived the Scopes trial, the contortions imposed on the definition of the fundamentalist movement would doubtless have played out differently. As it was, Bryan's untimely death a few days after the trial prevented him from exerting his influence and popularity to maintain wider boundaries for fundamentalism following the trial.[35]

Like Bryan, J. Gresham Machen carefully defined his relationship to fundamentalism. Machen was a prominent Presbyterian theologian who agonized over his relationship to the fundamentalist movement. In public, Machen eschewed the label even while he accepted a reputation as the intellectual leader of the movement. This careful boundary work on Machen's part was often ignored during the controversies of the 1920s. Friends and foes, and even Machen himself, generally assumed that he could speak for the fundamentalist movement.

Machen had reason to be cautious about identifying too closely with fundamentalism. He certainly did not fit the liberal stereotype of the wild-eyed, backcountry fundamentalist preacher. Machen came from an effete background and was educated at Johns Hopkins, Princeton Seminary, Princeton University, and the German universities of Marburg and Göttingen.[36] Socially, he was more comfortable enjoying a good cigar, a social drink, and abstruse discussions of culture, religion, and politics than attending a fire-and-brimstone tent revival. Theologically as well, he differed from many of the leading fundamentalists of his age in that he never accepted the theology of premillennial dispensationalism. Politically, he vehemently opposed efforts to legally require Prohibition on the grounds that it involved the state too much in private affairs.[37] Unlike many of the fundamentalist leaders and grassroots activists of the 1920s, he did not make struggles for control of American culture and education his primary concern. In the words of biographer D. G. Hart, Machen's primary interest lay in "preserv[ing] Presbyterian theology and church practice, and [he] limited his efforts against liberalism to the ecclesiastical sphere."[38]

Nevertheless, fundamentalists and liberals alike considered him the leading intellectual of the movement. His *Christianity and Liberalism* (1923) received a hearty welcome into the fundamentalist canon.[39] In addition, Machen willingly served during the early 1920s as the voice of intellectual fundamentalism. In 1924, he debated the negative position on the question "Does Fundamentalism Obstruct Social Progress?" with the liberal New York Old Testament scholar Charles F. Fagnani. In his argument, Machen identified with fundamentalism "in the broad, popular sense of the word." That is, he did not join specific fundamentalist organizations such as Riley's World's Christian Fundamentals Association. Nor did he ascribe to the theology of premillennialism. But he agreed that the broader fundamentalist movement represented more than these elements. In its broadest sense, Machen argued, the fundamentalist movement included him and all conservative evangelicals who had committed themselves to "checking the spiritual decadence of our age."[40] In public, Machen insisted that he found the fundamentalist label "distasteful" but he simultaneously equated fundamentalism with his preferred labels, "conservative" and "evangelical."[41] Privately, he told fundamentalist leaders that he considered them his spiritual "brethren"[42] and thanked them for their continuing "fellowship."[43] As had Bryan, Machen sought to construct for himself a unique position along the outer boundaries of the fundamentalist movement. He explicitly repudiated the label, yet in practice he allowed himself and others to view him as an intellectual leader of the movement.

The founders of the Evangelical Theological College in Dallas waged a different kind of struggle with the meanings of fundamentalism. Although they hoped during the 1920s to create an entirely new kind of fundamentalist seminary in the United States, they disagreed vehemently with other leaders, especially Norris and Riley, about what it meant to be a fundamentalist. As they created an independent fundamentalist educational institution, they aggressively challenged the right of Riley and Norris to unilaterally define the new movement.

Ironically, the original impetus for the new seminary came from Riley himself and the early leadership of his World Christian Fundamentals Association. Riley had called for the opening of a "Fundamentals Seminary" since the early 1920s.[44] Ultimately, fellow WCFA founders A. B. Winchester, Arno Gaebelein, Leander S. Keyser, W. P. White, W. H. Griffith Thomas, and Lewis Sperry Chafer took the initiative. All agreed that an entirely new kind of seminary was needed to combat the liberal seminaries that they believed were "shot through with a modernized pagan philosophy." Even the conservative seminaries, they feared, had drifted too far from the theology of dispensational premillennialism.

Bible institutes were not the answer, because they were not generally academically rigorous enough to produce pastors capable of high-quality theological work. No, in order for the dream of a worldwide network of well-trained fundamentalist pastors to be realized, the founders wanted a new type of school.[45]

Like many other fundamentalist school founders in the 1920s, the Evangelical Theological College founders sought to guarantee lasting doctrinal purity with an ironclad faculty creed. Faculty members had to sign this creed annually to prevent heresy and thus avoid the harsh controversies that plagued denominational colleges.[46] The Evangelical Theological College, which changed its name to the Dallas Theological Seminary in 1936, had some remarkable success in achieving its goals.[47] By 1926, the school needed to move to larger buildings to accommodate its rapid growth.[48] Student enrollment climbed quickly, from sixteen students in its first academic year (1924–1925), to twenty-seven the next year, to fifty students by the end of the decade.[49] By most standards, this was still a small school, but its influence grew rapidly. By the last quarter of the twentieth century, "Dallas" had become America's most influential seminary for training pastors in the theology of dispensational premillennialism.[50]

In spite of this success, the founders of the Dallas seminary had a very uneasy relationship with their fellow fundamentalists during the first years of the seminary's existence. Lewis Sperry Chafer, the energetic leader of the new seminary, worried about any association with Riley, Norris, and other prominent fundamentalist leaders such as Pennsylvania-based Methodist evangelist Dr. L. W. Munhall, and Tom Horton of the Bible Institute of Los Angeles. Chafer distrusted their motives, and he warned one correspondent, "Just what these four plunging men will do before they are checked remains to be seen."[51] Cofounder Arno Gaebelein agreed "it would be too bad if the new school was linked in any way with Fundamentalism."[52] These prescient leaders feared exactly what came to pass. Chafer worried about the "embarrassment" that might accrue to the meaning of fundamentalism if it became limited to the anti-intellectual, hyperaggressive, factious style of leaders such as Norris, especially due to the ways those belligerent leaders personified fundamentalism "in the public mind." Chafer worried privately that Norris's style would cause "the name 'Fundamentalist' [to be] utterly ruined, if it is not already."[53]

Nevertheless, the founders of the Dallas seminary still considered themselves fundamentalists "in the accurate meaning of that name."[54] And, in spite of their disagreements with Riley and Norris over the meanings of fundamentalism, the Dallas leaders continued to publicly identify themselves as fundamentalists in the early years of the decade.[55] However, once the school controversies brought intense pressure to bear on the fundamentalist image, most of the founders of the Dallas school distanced themselves from the newly restricted boundaries of fundamentalism. Although they had hoped to create a space for themselves as the real fundamentalists, the results of the school disputes forced them to retreat from fundamentalism by the end of the decade.

Lewis Sperry Chafer and other conflicted fundamentalists rarely let their definitional distress impede their educational activism. In Dallas, Chafer and his colleagues continued to build their premillennial seminary even as they disputed the propriety of a fundamentalist label. Many other conservative Protestants engaged in similarly dogged public activism. Many conservatives, including members of the booming Pentecostal movement, the revived Ku Klux Klan, Seventh-day Adventists, conservative Lutherans, and even some Catholics joined fundamentalist-led school campaigns. Many denominational conservatives, especially Presbyterians and Baptists, may not have considered themselves members of the fundamentalist wing of their denominations yet they also participated in mandating traditional Protestant teachings in their local public schools. Fundamentalists themselves often had ambivalent attitudes toward these allies. Many fundamentalists fought bitterly to define themselves in clear distinction to other Protestant groups. Yet in their school campaigns fundamentalists often welcomed and even courted support from a wide spectrum of conservatives. The rest of this chapter will describe both fundamentalist attempts to create workable boundaries with their conservative allies and the ways those meticulously constructed distinctions often broke down in practice.

Since their beginning around the year 1900, Pentecostal churches had achieved impressive growth. They offered a twofold appeal: a traditional Bible-based faith coupled with exciting, emotional worship services. Most Pentecostal groups believed just as strongly in an inerrant Bible as fundamentalists did, but Pentecostalism distinguished itself by its emphasis on an experience Pentecostals called the baptism of the Holy Spirit, which included the gifts of speaking in tongues and miraculous divine healing.[56] This new style of worship had caused debate among evangelical Protestants almost from Pentecostalism's inception. Some evangelicals embraced the Pentecostal message, while others attacked it.[57]

Newly minted fundamentalists in the 1920s continued this uncertain relationship with Pentecostalism. Norris drew attention to the "nervous temperaments" of many Pentecostals that "deceive[d them] into thinking that they [had] an extra degree of spiritual power."[58] In spite of this kind of criticism, many Pentecostal preachers achieved startling popularity. For example, Aimee Semple McPherson, one of the most prominent preachers associated with Pentecostalism, was said to have drawn bigger crowds than any itinerant entertainer in American history, including P. T. Barnum and Harry Houdini.[59] Some fundamentalists were attracted to this kind of evangelical success. The *Moody Bible Institute Monthly* described Pentecostals as "sincere Christians" and "well meaning persons."[60] Indeed, before McPherson's reputation had been tarnished by her scandalous six-day disappearance with a male associate, Gray argued that "Mrs. McPherson's personality, her sincerity and modesty [were] beyond criticism."[61]

Long before then, however, many self-identified fundamentalists had denied the right of Pentecostals to consider themselves part of the fundamentalist movement. One typical fundamentalist writer dismissed Pentecostalism as a kind of "hysterical fanaticism," arguing that "disorderly confusion in the assembly is not of God."[62] Another averred: "Usually people carried away by this movement are of a nervous, mystical, hysterical temperament, such as are considered a bit queer."[63] One critic worried that the alleged Pentecostal focus on emptying one's consciousness could lead to disastrous results: "A body," he warned, "detached from one's volitional control is ready for any counterfeiting work of the Devil or the flesh." Further, he argued that "outbreakings of sensuality have marred the history of the Pentecostal movement from time to time."[64] Another harsh fundamentalist critic argued that Pentecostals offered "delusive teachings" that made it a "dangerous movement which must work disaster in the church of God, because it is an imitation."[65]

However, some fundamentalists had trouble accepting this boundary as part of their own definition of fundamentalism. One controversial issue was the question of divine healing. Some fundamentalists retained the traditional Protestant belief that miraculous healing had ceased with the end of the Age of the Apostles.[66] Other fundamentalists, however, could not exclude divine healing from their self-definition as fundamentalists.[67] The attempt to define a movement that might or might not exclude miraculous healing, but somehow did exclude Pentecostalism, taxed the limits of early fundamentalists' ability to make sense of the boundaries of their new movement.

Fundamentalists in the early 1920s struggled to establish their relationships with other conservative Protestant groups as well. In most cases, this was a difficult and acrimonious process on both sides. As fundamentalists such as Riley, Norris, and Gray struggled to understand what it meant to accept a fundamentalist label, other groups waged similar battles. But at least one group of conservative Protestants experienced a relatively easy time defining their own stance toward the new term before the publicity of the Scopes trial. Missouri Synod Lutherans shared much of the conservative Protestant theology of fundamentalism. However, by the mid-1920s, Lutheran leaders asserted an identity clearly and definitively separate from "fundamentalism." In 1924, Missouri Synod Lutheran leader John Theodore Mueller pointed out that "after all has been said, there remains a sharp difference between Calvinistic Fundamentalism and confessional Lutheranism—a difference not in degree, but in kind."[68] Instead of bitterly debating this distinction, as they did with other groups, fundamentalists cordially accepted the status of Missouri Synod Lutherans as friendly neighbors and allies in the fight to maintain traditional Protestant doctrine in public schools.[69]

The process of definition and distinction became much more heated and angry when the borders were less clearly agreed on. Seventh-day Adventists shared much of the conservative, Bible-based Protestant theology of the fundamentalists. Yet in spite of such commonalities, some Seventh-day Adventists insisted on distinguishing themselves from fundamentalism. Fundamentalist writers and editors reciprocated by denouncing Seventh-day Adventism as a whole. In an article lumping Seventh-day Adventism in with other "Modern Cults," a writer in *The King's Business* noted Adventists' belief in a "false prophetess" with alarm. This writer also warned that Adventists should be considered a "great menace" because of their ability to lure away conservative evangelical believers.[70] Another contributor to *The King's Business* warned, "Seventh-Day Adventists, Who Teach much that is True, are Active. Look out for the False!"[71] The editors of the *Moody Bible Institute Monthly* warned readers emphatically that Adventist theology was riddled with "grave doctrinal errors."[72] It was not just that Adventist teachings were "a perversion of Scripture and blasphemy against God," Gray admonished, but they were close enough to the fundamentalist truth that earnest believers could be led astray. In this case, the hope for a clear boundary between fundamentalism and Seventh-day Adventism was continually frustrated by ideas and individuals that straddled any asserted boundaries.[73]

The relationship between fundamentalism and other neighbors, such as the resurgent Ku Klux Klan, was also hard to define clearly. Throughout the 1920s, some fundamentalists supported the Protestant activism of the Klan, while others disavowed it. Several voices in the MBI community defended the Klan, such as the Pennsylvania pastor who challenged his fellow fundamentalists to "investigate the Klan. So far I have found that the churches never had a more active ally, the state a more determined champion; our homes a more resolute defender, and lawlessness and vice a more powerful foe than the Ku Klux Klan."[74] In the end, however, Gray agreed with one Texas pastor who argued that "the great principle of Christianity is love. The outstanding principle of Ku Kluxism is hatred."[75]

In spite of this decision at the MBI, other fundamentalists, notably Norris, continued to give the Klan their earnest support. Norris ran advertisements announcing Klan functions and activities, and wrote sympathetic editorials defending the Klan's side of local controversies. In 1924, he went further than most other fundamentalists when he threw all of his prestige behind the Klan-led campaign to oust Texas Governor Jim Ferguson and to defeat the gubernatorial campaign of Ferguson's wife Miriam. Throughout the decade, some fundamentalists, like Norris, supported the Klan. Others opposed it fiercely. Many more muddled in the middle, unsure if they should support the secretive group, but unwilling to oppose fellow conservative Protestants.[76]

This expansive, confused middle allowed conservatives in the 1920s to operate in the public sphere without first having to tease out exact meanings for their labels. For instance, rank-and-file members of the revived 1920s Ku Klux Klan fought against the teaching of evolution in public schools. They exerted their considerable political power to mandate the readings of the Bible in public schools. Klan members did so without any apparent agony over whether or not such activism constituted "fundamentalism."[77] As we have seen, many leading fundamentalists disavowed the Klan. Certainly, since the Klan welcomed many theologically moderate Protestants, they did not fit the standard definition of the movement. Nevertheless, any distinction between the Klan and fundamentalism often vanished when it came to school activism.

Similarly, Pentecostals did not always succeed in building clear and impermeable boundaries with the fundamentalist movement. Some had no interest in calling themselves fundamentalists. They were just as certain as fundamentalists of their own superior knowledge of God's will. However, Pentecostals were no more monolithic than fundamentalists, and many had differing attitudes toward what historian Grant Wacker calls their "radical evangelical . . . cousins."[78] In some cases, Pentecostals hurled abuse at conservative evangelists such as Billy Sunday and Sam Jones. In other cases, though, Pentecostal periodicals reprinted the work of fundamentalists such as Gray.[79] Many also cherished the work of evangelists like Reuben A. Torrey, who had served as dean of both the Moody Bible Institute and Biola.[80] Further, at least one prominent Pentecostal denomination, the Assemblies of God, continued to explicitly identify itself as "fundamentalist" until at least 1928.[81]

In the same vein, some prominent Seventh-day Adventists, despite possessing an extrabiblical source of revelation in the writings of the prophetess Ellen G. White, referred to themselves as fundamentalists. Seventh-day Adventists, like fundamentalists, believed in the imminent, bodily return of Christ, and the plenary inspiration of Scripture. Furthermore, no prominent fundamentalists disputed the hostile appraisal of the editor of *Science*, who called Seventh-day Adventist George McCready Price "the principal scientific authority of the Fundamentalists." Price and his writings continued throughout the decade to play a leading role in fundamentalist educational thinking, regardless of Price's denominational affiliation.[82]

Some conservative Protestant groups developed an even more complex relationship during the 1920s with "fundamentalism." Holiness Wesleyans, including members of such groups as the Salvation Army and of denominations such as the Nazarenes and the Church of God (Anderson, Indiana), often assumed they formed part of the fundamentalist coalition. For instance, during the first years of the 1920s, when the fundamentalist identity of many groups was still uncertain, writers in the Church of God's *Gospel Trumpet* magazine explicitly identified themselves as "fundamentalists."[83] On the other hand, by mid-1924, a *Gospel Trumpet* writer could denounce both sides of the public controversies: "The secular newspaper and religious press teems with the great battle now on between the so-called 'modernists' and the so-called 'fundamentalists.' The strange thing about it is that so much error, as a rule, enters on both sides that it is hard to determine which is going the right direction."[84]

During this time of contested definitions, even many Catholics identified themselves as fundamentalists. As Maynard Shipley, the inveterate foe of the fundamentalist movement noted, there were a number of Catholics who considered themselves "of Fundamentalist persuasion."[85] A survey at the University of Michigan in 1925 found that almost 20 percent of Catholic students identified themselves as fundamentalists.[86] Clearly, these young Catholics did not consider themselves to be militant Protestants. Instead, they likely latched onto the widespread use of the term to imply a wider conservative impulse. Like the Catholic students, many Americans in the 1920s considered fundamentalism not as a narrow theological movement, but as a wider trend toward militant traditionalism.

Even denominational conservatives who fought against fundamentalists in denominational disputes often acted in tandem with fundamentalists on matters of education policy. For example, leading Texas fundamentalist Norris earned the condemnation of his fellow Texas Baptists in 1922. In that year, the Baptist General Convention of Texas (BGCT) condemned Norris's fundamentalist activism as "divisive, self-centered, autocratic, hypercritical and non-cooperative." The BGCT also criticized Norris for attempting to lead Texas Baptists out of the denomination and into a fundamentalist rump group, the Baptist Bible Union.[87] Nevertheless, the same convention also resolved against the teaching of evolution in Texas Baptist schools. Among Northern Presbyterians and Baptists as well, many conservatives who aligned against fundamentalists in denominational issues could still fight for traditional Protestant domination of schools.[88]

In practice, especially in public campaigns to save American schools, any attempt to explain away this tangled muddle of movements, alignments, and definitions confuses more than it clarifies. In spite of earnest attempts by leading fundamentalists to assert a simple definition or creed to identify "real" fundamentalists, such simple definitions do not shed much light on fundamentalist school activism. For instance, was Bryan not a fundamentalist because he avoided the term even though all contemporaries explicitly considered him the public face of the movement? Did Pentecostal activists act as "fundamentalists" when they fought against evolution in schools, but as "Pentecostals" when they went to church?

In the end, even if we could answer these questions, the answers would not help much to explain fundamentalist school campaigns. Instead, we need to acknowledge that fundamentalism, like other labels, had boundaries that were constantly under construction. Some contemporaries, such as the fundamentalist Catholic students at the University of Michigan, used the term to refer to all militantly conservative Christians. Others, such as Riley, attempted to define the term unilaterally.

But when Pentecostal leaders of the Assemblies of God denomination asserted that they were "fundamentalists to a man," they may have attached a variety of interconnected meanings to that self-identification.[89] They likely did not mean that they agreed with Riley's nine-point creedal definition of the movement, but they aggressively combated any nontraditional, non-Protestant teachings in schools, and they firmly believed in the truth of the Bible.[90]

Similarly, when Professor Charles R. Erdman of Princeton Seminary fought against fundamentalist domination of his Northern Presbyterian denomination, it did not mean that he wanted evolution to be taught in his local schools. It did not mean that he thought public schools ought to be pluralistic forums for the education of self-directed, skeptical, inquiring minds. Rather, it meant that he disputed fundamentalist direction of his denominational leadership. In Erdman's case, he was as conservative theologically as many leading fundamentalists, but he did not want to see his denomination torn apart.[91]

As fundamentalists moved to college campuses and state legislatures to fight for their vision of proper schooling, these tangled and contested definitions moved with them. Many of the supporters of fundamentalist school policies did not measure up to the definitions of the movement asserted by leaders such as Riley. Yet those supposed nonfundamentalists ardently supported traditional Protestant teaching in America's public schools and they sometimes considered themselves fundamentalists when they did so. Moreover, as fundamentalists struggled to assert control of those schools, the meanings of their movement endured inexorable pressure to conform to a stereotype that many fundamentalists found unacceptable. Activists who may have called themselves fundamentalists in 1924 may not have done so in 1929. These changes in meaning, self-identification, and the boundaries of fundamentalism not only changed fundamentalism, but they also contributed to the successes and failures of fundamentalist school policies nationwide.

PART II

God and School

CHAPTER 3

Campus Skirmishes

That a profound cultural clash should be contested as a matter of school policy should not come as a surprise. As historian Lawrence Cremin has pointed out, "Many of the great twentieth-century battles over traditionalism and modernity . . . were ultimately framed as educational issues."[1] Other historians agree. James L. Axtell has called education the "most sensitive instrument" for noting cultural change.[2] Although, as Benjamin Justice has argued, this "warfare thesis" has often overemphasized conflict, in the case of fundamentalist educational policy, warfare has certainly been the norm. Just as in more recent "culture wars," the struggles for schools in the 1920s were really struggles for the soul of America itself.[3]

For most fundamentalists, public activism began with concern over the teaching at America's colleges and universities. Many activists on both sides used a loosely defined fear of evolutionary theory as a symbol and rallying cry for these battles. In fact, although the teaching of evolution on college campuses did form one important issue, the more fundamental struggle took place over the role of Protestant belief at those schools. Just as they would in their struggles at elementary and secondary schools, fundamentalists fought for much more than simply banning the teaching of evolution from colleges. In short, fundamentalists battled to maintain the traditional hegemony of Protestant belief. They struggled to ensure that college students could still earn a degree without accepting the intellectual presumptions of theological modernism and philosophical materialism.

In spite of their dismay at what they perceived as the atheistic results of teaching evolution, several leading fundamentalists insisted that evolutionary theory could and should be taught, especially at the collegiate level. Bryan, for instance, repeatedly and adamantly insisted that evolution must only be banned when taught as a fact.[4] Fundamentalist educator Alfred Fairhurst suggested that "both sides"—organic evolution and the Genesis account—be taught in universities.[5] Evolutionary theory was entirely improper for young students, Fairhurst insisted. But he stipulated that in higher education "it ought to be taught *honestly* and fully to the select few who have the ability to comprehend it in all its bearings."[6] Even the antievolution firebrand T. T. Martin agreed that some students needed to be exposed to evolutionary theory. Martin suggested a new set of "graded books, from primary to university," each of which could present "fairly and honestly both sides of the Evolution issue."[7] James M. Gray of the Moody Bible Institute also maintained that students should be exposed to evolutionary theory, under proper conditions. It should not be taught as fact, nor should it be taught to very young students. Nevertheless, Gray insisted throughout the 1920s that teaching materialistic evolution as one scientific theory was not inherently antithetical to fundamentalist belief.[8]

When fundamentalists struggled for control of higher education, these nuances of position were usually drowned out by heated rhetoric on both sides. As would happen with later debates over teaching at elementary and secondary schools, most of these collegiate controversies became known most prominently as fights over the teaching of evolution. Yet fundamentalists repeatedly insisted that evolution was only one example of a conglomeration of ideas that led students away from their faith. They fought to maintain campuses safe for evangelical belief. At the most prestigious colleges and the big public universities, fundamentalists usually lost. In this first round of the twentieth-century culture wars, fundamentalists had to concede that the mainstream intellectual life of leading colleges had been ineluctably won over to a more secular pluralism. However, that did not mean that they gave up their dream of fundamentalist higher education. Instead, fundamentalists realized in the early years of the 1920s that their grand expectations of reestablishing cultural and theological control of leading colleges and universities were not to be. They would need to develop their own alternate network of higher education free from the dominant intellectual beliefs that had taken over mainstream higher education in the early years of the twentieth century.

By 1920, conservative Protestants were familiar with the charge that German-trained university professors had corrupted America's colleges. As fundamentalist activist William Bell Riley exhorted his readers, the first step toward saving colleges was to rescue them from an "avalanche of German rationalism."[9] Antievolution evangelist Martin reminded an audience in Los Angeles that Americans themselves were to blame for this alarming state of affairs. After all, he scolded, "we sent our young men to the great German universities, and, when they came back, saturated with Evolution, we made them Presidents and head-professors of our colleges and great universities."[10]

Although Martin and Riley were masters of hyperbole, they were entirely correct in this case. The shift on many of America's campuses away from an explicitly religious and specifically Protestant approach to scholarship was largely due to the earnest work of a late-nineteenth-century crop of faculty and administrators trained in German graduate schools. This new generation—unlike their predecessors—made it their mission to defend unpopular and controversial thinking among their faculty, treasuring academic freedom as a higher value than traditional Protestant doctrine. As historian Merle Curti noted, "When the German-trained thinkers replaced the more orthodox theologians in American colleges and universities the league of Christianity and higher education was severely shaken."[11] By the 1920s, argued George Marsden, American universities had been transformed from denominational institutions into schools of "established nonbelief."[12]

Fundamentalists were appalled by this widespread change. Instead of a faculty dedicated first and foremost to the promotion of the traditional truths of Protestant Christianity, such as the virgin birth and deity of Christ, the inerrancy of the Bible, and Christ's pending return, some members of college faculties debunked these ideas. By 1909, when *Cosmopolitan* writer Harold Bolce ventured onto several college campuses to survey their teaching practices, he warned that readers would be "astonished" by his report. Readers of the popular series learned that professors no longer respected the Bible, the Declaration of Independence, or the traditional family.[13]

Compounding this change in the tenor of academic life was the rapid growth of the role of colleges and universities. Between 1871 and 1931, the overall population of the United States grew from almost 40 million to over 123 million people. The growth in the numbers of colleges easily kept pace. In 1871, there were 563 colleges in the country. By 1931 that number had grown to 1,460. More important, the proportion of Americans earning degrees from those colleges far outstripped the average population growth. In 1931, 7.4 percent of Americans between the ages of seventeen and twenty-four were enrolled in a college program, whereas in 1869 only 1.3 percent of Americans in that age bracket had done so, and the number of students who remained in school long enough to earn degrees increased equally remarkably. In 1871, American colleges and universities awarded 7,852 undergraduate degrees. By 1931, that number leaped to 138,063. The number of terminal degrees awarded increased by an even greater proportion. While only fourteen PhDs were awarded in 1871, 2,654 were awarded in 1931. Clearly, the numbers of people exposed to collegiate life and training had increased remarkably in this short time period. Along with the profound change in the way many college professors and presidents viewed their cultural role, this explosion in the numbers of people attending higher educational institutions meant that many more Americans would be concerned with what was taught there.[14]

In leading schools, these fights began soon after the Civil War. One widely publicized early controversy arose between the traditionalist president of Yale, Noah Porter, and the famous scholar William Graham Sumner. In 1869, Porter objected to Sumner's use of an "anti-Christian" text written by the prominent evolutionist Herbert Spencer. Although Porter boldly announced that he would not allow "atheism or anti-Christianity be taught in any of [Yale's] chairs, either directly or indirectly," he recognized that he ultimately had to respect Sumner's choice of texts. This recognition signaled the shift toward a more pluralistic atmosphere on college campuses throughout the nation.[15]

The changes, however, did not happen overnight. Many top administrators at leading schools bucked the trend toward secularization. James Burrill Angell at the University of Michigan is one example of the complex ways these changes were implemented during the late nineteenth century. In some ways, Angell led the University of Michigan away from an established Protestant tradition. For instance, he made attendance at chapel services voluntary in the 1880s. However, Angell envisioned a state university that would promote no single denomination but would actively encourage evangelical Christianity. Even as he changed the university to stay in step with the secularizing times, he also actively sought out Christian faculty and encouraged denominations to work among the Michigan students. He also opened up university facilities to Protestant groups such as the YMCA, YWCA, and to evangelist Dwight L. Moody.[16]

Other schools took Angell's evangelical approach even further. Many remained true to the ideals of the traditional Protestant college. The most prominent example was Wheaton College in Illinois. Under the leadership of Charles A. Blanchard, the liberal arts college remained committed to a theological and educational approach that had been common during the nineteenth century. By 1913, Blanchard had personally accepted the theology of dispensational premillennialism. This theological commitment pushed Blanchard to resist the changes taking place at many other colleges and universities. By 1916 he had devoted himself and his career to the conservative side of the coming modernist-fundamentalist controversies. During his lifetime, the force of his personality and his energetic leadership were enough to ensure that Wheaton College became a bastion of fundamentalist higher education. Only after his death in 1926 did the board of trustees adopt a nine-point creed to ensure continuing fidelity to the doctrines of fundamentalism.[17]

In addition, some individual scholars avoided the stereotypical conversion to the morality of academic freedom and the preeminent value of skeptical inquiry. Reuben A. Torrey, for example, was a Yale-educated pastor and evangelist who became dean of both the Moody Bible Institute and the Bible Institute of Los Angeles. Like many of his contemporaries, Torrey spent a year (1882) studying in German universities. Instead of converting him to rationalism and the value of academic freedom, this study year convinced Torrey of the centrality of the idea that the Bible must be accepted as the inerrant Word of God. For Torrey, graduate study in Germany committed him to the theology that would make him one of the godfathers of the fundamentalist movement of the 1920s.[18]

In addition, conservative factions had some success in guarding key positions. In 1884, for example, conservative Presbyterians successfully ousted Professor James Woodrow from the Columbia Seminary in South Carolina for his unorthodox views on evolution.[19] Other early victories were pyrrhic. For example, although conservative Methodists were finally able to eliminate liberal Old Testament professor Hinkley G. Mitchell from Boston University, they—and their fundamentalist successors—lost in the long run. After the loss in the Hinckley case, liberal members of the Methodist Episcopal Church North ruled that bishops no longer had theological oversight of colleges. This move opened the way for even more unorthodox religious teaching at denominational schools.[20]

Unorthodox teaching, both at denominational colleges and at public universities, often acted as the catalyst that spurred fundamentalists into public controversies with liberals and theological modernists. Fundamentalists loudly complained that students were being indoctrinated away from their evangelical faith. One early target of conservative ire was the bastion of theological modernism, the University of Chicago. Under the leadership of William Rainey Harper, who served as president from 1891 to 1906, the Baptist-founded school soon became the headquarters of modernist thought. Conservative Protestants noticed this trend with alarm. In 1906, for instance, New York Baptist pastor and future fundamentalist leader John Roach Straton attacked the "alleged scholars" at Chicago for abusing the "prestige of their position" to destroy the Christian faith of their students.[21] Other future fundamentalists reported that the University of Chicago "wrecked" the faith of every student who attended.[22] Conservative pastors complained that children left home as good Christians and returned as "avowed infidel[s]."[23] Riley, founder of the World's Christian Fundamentals Association (WCFA), warned that the divinity schools were no better than the other academic programs. Instead of training pastors in the traditional faith of evangelical Protestantism, many seminaries, especially the University of Chicago's Divinity School, had become "hot-beds of skepticism."[24]

Nor were public schools safe from fundamentalist attack. As early as 1917, Riley accused state universities of having turned into "Hot-Bed[s] of Heterodoxy."[25] This was a common charge. At the founding of Riley's World's Christian Fundamentals Association in 1919, the committee in charge of investigating the state of higher education grimly reported to the assembly that "modernism [had] captured very many of our schools" and warned of the "grave danger of sending [students] into the skeptical atmosphere." By 1919, one speaker warned, this battle between orthodoxy and theological modernism had already been fought, and most state colleges had already been taken over by atheism and anti-Christian teaching.[26]

Fundamentalists such as Martin described in graphic detail the results of student exposure to the ideas of evolution and atheism when they were away at public universities. Martin explained that students of college age were "romantic," and easily influenced by the latest fad. To their dismay, students later would discover that evolutionary thinking led them to atheism. Martin quoted one former student's complaint: "My soul is a starving skeleton; my heart a petrified rock; my mind is poisoned. . . . I wish I had never been to college."[27]

Fundamentalists were motivated by more than just gruesome anecdotal evidence. Many reacted with alarm to James H. Leuba's 1916 study *The Belief in God and Immortality*, which suggested that the teaching on America's college campuses had become hostile to evangelical belief. Bryan, for instance, often cited this study as evidence that college teaching had become pernicious. If eighty-five percent of college freshmen described themselves as believers, but only fifty to fifty-five percent of graduates did, Bryan reasoned, then the teaching going on at America's colleges must somehow promote disbelief.[28]

Many fundamentalists reacted to Leuba's work with similar outrage, and several prominent fundamentalists conducted their own surveys to understand college culture. In 1919, Wheaton College President Charles A. Blanchard conducted a survey of fifty-four schools throughout the Midwest and Plains states. He was determined to find out which colleges were teaching heterodox ideas and which allowed students too much license. He asked school officials a series of questions, including whether they promoted evangelical Protestantism, whether the faculty taught evolution, whether they considered the Bible "the inspired Word of God," whether students and faculty were allowed to dance, drink, or play cards, and finally, whether the college allowed "secret associations" such as fraternities or sororities.[29] Blanchard found his conclusions alarming. He reported to the 1919 founding meeting of WCFA that most colleges had slipped into teaching modernism and atheism even though those who were surveyed called themselves Christian. Thus, it was impossible for students and parents to safely choose a Christian college that had remained true to the doctrines of conservative evangelical Protestantism. Only rigorous action, Blanchard concluded, could save students from the prevailing atheism taught even at nominally Christian colleges.[30]

Blanchard was the most prominent but not the only conservative Protestant to conduct this kind of college survey. Fundamentalist G. W. McPherson published his survey of 211 "leading educators" in 1919 as *The Crisis in Church and College*. His results were similar to Blanchard's. He estimated that there were two million students currently at risk of being led away from true Christianity. Like Blanchard, McPherson suggested that only an aggressive, confrontational campaign could save "the flower and the hope of American Christianity."[31]

Alfred Fairhurst, former college professor and educational activist, joined the field of fundamentalist surveys with his 1921 book *Atheism in Our Universities*. Fairhurst had sent questionnaires to prominent educators, including presidents and ex-presidents of universities, superintendents of state school systems, and normal school professors. Fairhurst wanted to know to what extent and for what purpose evolutionary theory was taught. Most of the respondents supported the teaching of evolution in higher education, but some agreed with Fairhurst that the teaching should be restricted to older students. Charles W. Eliot, former president of Harvard University, explained that the teaching of evolution could be dangerous to the faith of younger students and should be restricted to "older pupils."[32]

Fairhurst's primary interest was not in science, however, but in theology. He believed that widespread evolutionary thinking must lead to widespread atheism. In addition to his alarm at the prominence of evolution on college campuses, Fairhurst reported with consternation that atheism was a prominent and growing part of collegiate education. At the Ohio State University, for example, one correspondent reported that seventy-five percent of the professors were atheists and that only twelve percent were Christian believers. Further, Fairhurst included anecdotal evidence of the loss of faith among college students. He reported, for instance, that sixty percent of the students in a three-year biology course at Missouri State University were transformed from believing Christians to atheists.[33] Interestingly, this 1920 survey included three responses from state superintendents of education who reported no controversy about the teaching of evolution at the secondary or elementary levels. Thus, while the controversy had already begun to rage at college campuses, many informed elementary and secondary school educators still had no idea that a similar storm was brewing for them.[34]

In response to this alarming situation, many fundamentalists suggested plans to regain cultural control of higher education. Especially in the case of private denominational schools, fundamentalists argued that gaining control of the curriculum was a simple matter of snapping the purse strings tight. Until denominational schools returned to teaching ideas acceptable to the denomination, many fundamentalists argued, fundamentalists could refuse funding to heretical schools. Blanchard strongly supported this tactic. He used his survey data to compile two lists: the "safe" schools and those that had strayed. Only the theologically orthodox, he argued, should receive any denominational support.[35] Other prominent fundamentalist college surveyors agreed. McPherson suggested a simple solution: "When they dismiss all such [theological modernist] teachers and abandon their textbooks and reference books," McPherson promised, "then I shall give them a hearty support."[36] T. C. Horton, dean of the Bible Institute of Los Angeles called on the fundamentalist community to "deny the schools the privilege of taking our money and using it to destroy the faith of our children!"[37] Martin argued that the cultural war would be won quickly. Since denominational schools could quickly be brought into line with financial pressure, Martin felt that secondary and elementary schools must soon follow. After all, it was in college that teachers learned their trade, and Martin believed that teachers would soon be taught only fundamentalist doctrine.[38]

These rosy dreams would soon be contradicted by the realities of the protracted struggle over higher education. Throughout the 1920s, fundamentalists fought for control over both private denominational colleges and public universities. The situation was more complicated at the state schools, since fundamentalists could not pull denominational levers of control. Nevertheless, fundamentalists did not cede control over public curricula to the evolutionists, liberals, and secularists. Some of the most public controversies took place at the University of Wisconsin in Madison, the University of Minnesota in the Twin Cities, and the University of North Carolina at Chapel Hill. All of these schools were major research universities, and all ended up successfully defending their German-modeled academic freedom and rationalism. However, leaders at all three schools quickly realized that fundamentalists were willing and able to muster broad popular support for their cultural campaigns.

One of the first college presidents to tangle with fundamentalist activists was Edward A. Birge of the University of Wisconsin. Birge was himself religious, a Congregationalist deacon and a Sunday school teacher.[39] He was also a prominent zoologist, and his scientific training and background had convinced him of the validity of evolution. Birge believed in God as the prime mover of creation. For twenty-five years he had taught students that they need not dismiss Darwin and evolution in order to remain faithful Christians. This strident belief in theological evolution prompted Birge to begin a controversy with one of the most prominent antievolutionists of the day, William Jennings Bryan, a conflict Birge would later regret.

Bryan had been invited to give a lecture at the Madison campus on May 5, 1921. His speech was one he had been giving across the country, "Brother or Brute." In the talk, Bryan blasted Darwinism and accused it of lacking scientific merit. In spite of the applause, Birge, sitting on the platform with Bryan, listened in dismay. Afterward, Birge could not maintain a polite facade and accused Bryan of leading students to atheism himself. Birge asserted that it was Bryan, not Darwin, who forced students into a false choice between religion and science.[40]

This was exactly the kind of controversy Bryan had been looking for. An experienced politician and populist icon, Bryan happily seized on Birge's comments as a defense of evolution at America's public universities and rushed to mobilize popular support against such teaching. As would other fundamentalist leaders, Bryan used the rhetoric of "evolution" in complicated ways. In many cases, Bryan objected directly to the teaching of evolutionary theory. He argued that it had not been proven adequately and that to teach it was an abrogation of academic and scientific responsibility. But Bryan also used the teaching of evolution as a symbol of the degraded status of Protestant belief in mainstream intellectual culture. He demanded more of his opponent than an end to the teaching of evolution. He also insisted that Birge reaffirm the dominant role of traditional Protestant theology on campus.

In Bryan's attack, he blasted Wisconsin as a typically out-of-touch public university, taking public tax money to teach students evolution and atheism against their parents' wishes. He suggested a warning label that might be placed on the university's classrooms: "Our class rooms furnish an arena in which a brutish doctrine tears to pieces the religious faith of young men and young women; parents of the children are cordially invited to witness the spectacle."[41] Bryan soon laid out his demands: professors must stop teaching evolution, and President Birge must sign a statement affirming his own belief in the Genesis account of creation, the virgin birth, and biblical miracles.[42]

Birge quickly repented of his role in this contrived controversy, but he was still too hotheaded to allow Bryan the last word. He and Bryan engaged in a series of newspaper exchanges, in which Birge called Bryan "crazy,"[43] and Bryan called Birge an "autocrat."[44] Bryan even offered to pay Birge one hundred dollars if Birge would sign a note stating that he had descended from an ape. Birge refused.[45]

This first major public blowup at a state university over fundamentalist issues allowed each side to take the measure of the other. The administration of the University of Wisconsin remained unchanged. Professors remained free to teach scientific theories such as evolution. Perhaps more important, Birge did not endorse the intellectual hegemony of Protestant theology in university life, even symbolically. But President Birge learned to his surprise of the wide public support for traditional Protestant ideas on his campus. As a deeply religious man, a Protestant no less, he was confounded by the bitter accusations he faced of promoting atheism. At the first hint of controversy, he was deluged with letters, some long and argumentative, others short and threatening. As the controversy wound down, Birge warned Edwin Conklin, a prominent Princeton biologist and supporter of the teaching of evolution, about entering into controversy with Bryan. As soon as one becomes associated with evolution in the public mind, Birge wrote, "You will receive an enormous number of letters and much fool printed stuff." At the very least, Birge's experience had put school administrators on notice to expect vocal opposition to the teaching of evolution. It had also forced Birge away from the traditional role of university president as both religious and academic leader. Birge, as would other presidents of public universities, had to choose between the values of academic freedom and Protestant hegemony.[46]

The controversy in Madison also warned fundamentalists of the deeply entrenched nature of their liberal and secular opponents. As they feared, higher education had become largely the preserve of a pluralistic approach to intellectual life. But the spat also gave Bryan a taste of popular success in the fight against the teaching of evolution. He had tried out some of the arguments that soon became his antievolution stock-in-trade, including the idea that "the hand that writes the pay check rules the school."[47]

This skirmish at Wisconsin intensified a protracted struggle for fundamentalist activists. Of these school campaigns, the legislative drive to take cultural control of elementary and secondary schools garnered the most attention from many fundamentalists, as well as from most of their opponents. As the next chapter will describe, many of these battles became known as evolution fights, although the bills often included a much more comprehensive cultural agenda.

But even as the battles over legislative attempts to protect the evangelical faith of elementary and secondary school students began to attract fundamentalist energy and attention, fundamentalists continued to fight to retain traditional Protestant intellectual dominance at state colleges. The publicity of the Madison controversy soon encouraged collegiate controversy elsewhere. In October 1922, Bryan made two speeches in Minneapolis, Minnesota. Each one repeated his favorite themes: that public schools should teach doctrines in accord with traditional Protestant belief and that colleges have a moral obligation to be neutral toward the question of human evolution.[48]

Following the lead of professors from Madison, several professors from the University of Minnesota protested. Quickly picking up Bryan's side of the controversy, Riley organized the Minnesota Anti-Evolution League, which grabbed headlines by demanding that public schools, including colleges, abandon the teaching of evolution, or face an "appeal to the legislature for the enactment of such laws as shall eliminate from our tax supported school system this antiscientific and antiscriptural theory of man and the universe." Just as Bryan had, Riley used evolution as a rallying cry for a much wider campaign. Riley was irked not only by the teaching of evolution but by the embrace in Minneapolis schools of a clutch of "anti-Christian theories."[49]

Luckily for beleaguered University of Minnesota President Lotus Delta Coffman, Riley's attention was soon distracted by other college controversies, including one at the public college system of North Carolina. Riley was content, for the time being, to blast Coffman in the pages of Riley's magazine, to demand the removal of offensive textbooks—which Coffman forcefully refused to do—and to bluster that "this war will not abate until teachers and text books that scorn the Christian faith are gone!"[50]

In spite of such threats, however, the situation at the University of Minnesota quieted down for a while, but only because Riley and other fundamentalist activists converged on North Carolina to demand similar changes at its state schools. The particulars of the controversy in the Tar Heel State were unique, but both sides soon took up now-familiar positions. Both sides used evolution as a complicated symbol of a much broader intellectual trend. The spark that ignited the debate in North Carolina came in 1922 when the administration of the University of North Carolina at Chapel Hill invited evolutionist Charles Allen Dinsmore to give the prestigious McNair lectures. Fundamentalists and conservatives alike were outraged that the university could make such a controversial decision. Dinsmore outraged cultural and religious traditionalists with his assertions that religious revelation must play second fiddle to the findings of evolutionary science.[51] Building on this sense of popular discontent, fundamentalists such as Riley and Boston pastor Jasper C. Massee engaged in widely publicized debates with scientists from North Carolina State throughout 1922. The academic scientists argued for freedom of thought and the value of unfettered academic inquiry and teaching, while Riley and Massee hammered home the idea that public schools were responsible to the taxpaying public for their curricula. At issue was not merely the teaching of evolution, but the nature of knowledge. Massee blasted his foes for both their misplaced belief in evolution and their inability to recognize the shortcomings of their own intellectual presuppositions. Mainstream scientists, Massee charged, had become so caught up in their own intellectual hubris that they had made a fatal mistake. In his opinion, they failed to recognize the simple truth that mere humans "cannot by intellectual inquiry grasp an infinite God."[52]

The contests at the Universities of Wisconsin, North Carolina, and Minnesota were among the most publicized examples of the controversies that raged at state schools throughout the country during the years leading up to the Scopes trial. At the same time, fundamentalists were battling for control of denominational colleges and universities. It had been the leading role of modernists at the University of Chicago that had riled up many conservative Baptists and convinced them to join the fight for control of colleges in the first place. Like the personal and vicious debate among Baptists over the teaching at the University of Chicago, many of the fights over denominational schools showed a venom and a destructiveness not seen in the fights over public colleges. As with many family fights, it was because the two sides were relatively close that these debates often became intensely personal affairs.

In general, fundamentalists understandably felt a much greater right to insist on the doctrines taught at private schools than at public ones. As Baptist fundamentalist editor Curtis Lee Laws charged, "We who send our children away to Baptist schools do not wish them sent back with their faith shattered. . . . Unless our denominational schools prepare our boys and girls for the responsibility of life there is absolutely no reason why they should exist."[53] Just as in other college fights, the issues involved in the vitriolic fights for control of denominational schools expanded beyond the teaching of evolutionary theory. At denominational schools, the issues included not only evolution but also codes of conduct for students and faculty, choice of textbooks, and control over the personal theological and philosophical beliefs of teachers and students. For fundamentalists, some of these contests ended as victories in their first decade of struggle for control of colleges, but most were humiliating and dispiriting failures. As we have seen, many fundamentalists had brimmed with optimism at the start of the 1920s. In a few short years, however, they realized that most of their denominational schools, just like the large public universities, had been irretrievably won over to a dominant intellectual culture of pluralism and theological liberalism.

For some religious colleges, the theological and cultural controversies of the decade spurred their administrators to make their fundamentalist leanings more explicit. At Gordon College, in Massachusetts, the sense of controversy inspired the school's leaders to clarify its conservative theology. Beginning in 1922, faculty members were required to sign the school's conservative doctrinal statement of faith.[54] At the same time at Wheaton College in Illinois, President Blanchard was prompted by his sense of working in "this time of confusion and uncertainty" to interrogate his faculty, hoping to purge any theological modernists. Teachers were asked if they were Christian and whether they believed in the Genesis account of creation. They were also questioned about their personal habits: whether they smoked, danced, played cards, attended theaters or movies, or even "associate[d] with worldly people in other amusements." Finally, Blanchard required the faculty to sign an eight-point statement of faith that included such fundamentalist standards as an inerrant Bible, a Genesis creation, a divine Christ, His miraculous birth, life, and substitutionary sacrifice.[55]

This firm decision to identify themselves with the fundamentalist side of the debates at colleges was unusual. Most denominational schools experienced a split, with some members of their faculty, administration, and community supporting one side of the controversy or the other. Some of the most vitriolic fights occurred at Baptist schools such as Wake Forest in North Carolina and Baylor in Texas. However, other denominations were by no means immune.

At most of the schools, a common pattern emerged. Fundamentalists generally attacked the leaders of denominational schools, charged them with personal unbelief, and accused their administration of being sympathetic to evolution, atheism, and theological modernism. At schools in which theological modernism was in the ascendance, such as the University of Chicago, most of the fundamentalists' attacks were fruitless. However, at schools with a large body of sentiment sympathetic to fundamentalist claims, administrators were forced to demonstrate their own personal orthodoxy, as well as the orthodoxy of the entire faculty. Especially at some unendowed Baptist schools, fundamentalists wielded the powerful weapon of funding. If they could prove their antagonists had strayed into heterodoxy, conventions could vote to restrict their budgets. In many cases, administrators were forced to walk a tricky line between fundamentalism and liberalism in order to keep the budget solvent.

The long fight at Wake Forest University in North Carolina was a case in point. The president of the Baptist school, William Louis Poteat, came under fire for both his own theological beliefs and on the charge that he was sympathetic to the doctrine of evolution. In many ways, President Poteat embodied the new kind of college president fundamentalists feared. He had been educated in Germany and made no bones about his teaching of evolution.[56] It was Martin who opened the attack on Poteat in the very first days of the decade. In January 1920, Martin accused Poteat of believing in a "modernist" view of reconciliation, meaning that anyone who repented could be forgiven. According to Martin, this allowed "the Jew, the infidel, and the Mohammedan to go to heaven if only they quit sinning." Martin later accused Poteat of accepting evolution, leading Baptist Wake Forest students into atheism and away from real science. Citing the Baptist context, Martin called for Poteat's resignation on the charge that Poteat had accepted money under false pretences (i.e., Poteat had been employed to teach Baptist doctrine, according to Martin's charges, but he had failed to do so).[57]

President Poteat defended himself and his administration much differently than public university presidents had. Instead of defending a platform of academic freedom, Poteat defended his traditional Baptist values. In response to the accusations made by several itinerant fundamentalists who traveled throughout North Carolina, Poteat asserted that theological evolution meshed well with traditional Baptist doctrine. Further, Poteat sought to discredit the scientific claims of his accusers. In spite of repeated assertions of disapproval from local Baptist associations, Poteat was able to bring the state Baptist convention to his side in 1922. He accomplished this with a dramatic speech, New Testament in hand, in which he asserted his deep and traditional Christianity. However, this victory only papered over the deep divisions within the convention. Poteat was subjected in the coming years to repeated attacks by local and national fundamentalists.[58]

While Poteat was struggling to stay one step ahead of his fundamentalist foes, other denominational colleges experienced similar incidents. Riley opened the decade with an inquiry into his local Baptist school, Carleton College. Although Carleton was technically owned by the Congregational Church, it was operated by the Minnesota Northern Baptists, so Riley felt justified in complaining to the dean in 1921 that evolution was being taught in violation of Baptist orthodoxy.[59] When he did not receive a satisfactory response, Riley began his assault. He accused Carleton of doing nothing more than "White-Washing Infidelity," and he called it "modernistic without reserve." Riley even conducted some informal surveys of Carleton students, until he found one woman who agreed that Carleton's campus "was full of skepticism." For Riley, these were serious charges indeed. However, as with his attack on the

University of Minnesota, this fight was soon shelved until after the frenzy of the Scopes trial had come and gone.[60]

One of the first college controversies to distract Riley's attention from this menace in his own neighborhood was an inquisition at Southern Methodist University (SMU) in Dallas, Texas. Riley joined allies such as Norris and Horton in championing the accusations of Methodist evangelist W. E. Hawkins Jr. Hawkins had sniffed out unorthodox teaching at SMU and he was determined to purge the school of modernist teaching and evolution. One of Hawkins's complaints was the presence on the faculty of John A. Rice, whom fundamentalists accused of modernism. In the words of Norris, modernists like Rice had gone wrong when they studied "in Chicago University where they got the forty-second echo of some beer-guzzling German Professor of Rationalism."[61]

When the leadership at SMU denied the charges of modernism, as did the administrations of many denominational schools, Hawkins went on the offensive. He gathered together students from the school, as well as from Texas Woman's College and Southwestern University in Georgetown, Texas, and held a well-publicized interrogation. One former student, the child of missionary parents, complained that the campus was "teeming" with modernism.[62] Another student produced her notebooks, in which she had recorded the answers professors had given her to questions Hawkins had instructed her to ask. Not surprisingly, the answers she received were not acceptable to Hawkins. Was Jesus's birth divine? One professor answered: "I do not denounce the divinity of Christ, but doesn't all man contain a portion of divinity?"[63] To the ears of fundamentalists such as Hawkins, Norris, and Riley, this was proof enough that modernism had found a home in Texas. "I believe if I had continued to accept what was being taught to me," another student reported, "that my faith would not have held out very long."[64]

After reviewing this evidence, Horton concluded that this lamentable state of affairs was typical of denominational schools everywhere, as well as in state-funded colleges in which innocent students were injected with the "pernicious poison" of theological modernism and evolution.[65] Charles G. Trumbull, editor of the conservative *Sunday School Times*, agreed. The danger at denominational colleges and seminaries was even greater than at state schools, he said, because it came in the guise of Christian education. In Trumbull's opinion, this "insidious injection of poison" was a threat to civilization itself.[66]

Other early fundamentalists viewed the accusations against SMU with alarm. Lewis Sperry Chafer anticipated the break that would soon occur among fundamentalists. He worried that the trial would link the new fundamentalist movement to other religious inquisitions. Instead of a movement based on a theology with long roots in American and British intellectual history, Chafer worried that fundamentalism would become tainted with the stereotype of the xenophobic southern hillbilly.[67]

The trial at SMU also inspired one of the longest-running fundamentalist college controversies of the era. Shocked by the open avowals of theological modernism and evolution at the Methodist colleges, Norris vowed to investigate Baptist colleges in Texas. This began his long career as a gadfly to the administration of Baylor University.[68]

It did not take Norris long to stir up trouble at Baylor. His first charge against the school concerned a sociologist, Grove Samuel Dow. Norris dissected Dow's *Introduction to the Principles of Sociology* and found it chock-full of evolutionism. Although Dow attempted to defend himself by excising objectionable parts of the book and by arguing that he had never taught that humans had evolved from animals, Norris's charge stuck. Dow lost his faculty position, and the Baptist General Convention of Texas resolved that evolution would be banned from all Baptist colleges in Texas.[69]

This victory came at a high cost for Norris, however. In spite of the delight expressed by fellow fundamentalists such as Martin and Riley, Norris's conduct in the Baylor case earned him an unofficial censure at the hands of the Texas Baptist Convention. Fellow Texas Baptists were dismayed by the viciousness with which Norris made his charges and by his eagerness to publicly attack Baylor without first consulting with the college president, Samuel P. Brooks.[70] Some of the leadership of the premillennial Evangelical Theological College in Dallas (later Dallas Theological Seminary) denounced Norris's "angry polemics" in this campaign.[71] In the words of Dallas founder Arno Gaebelein, "The other side is going to use Mr. Norris and his reputation against the Fundamentalist Movement."[72] These fundamentalists had a reason to worry. In the years to come, the liberal enemies of fundamentalism would attempt to exaggerate Norris's reputation as a violent, demagogic thug to pigeonhole the entire movement.

However, none of these criticisms could slow down Norris, the "Texas Cyclone." He continued to blast President Brooks and selected members of the Baylor faculty. Norris often compared Brooks to President Poteat of Wake Forest in North Carolina. Both men, Norris argued, had to go. Norris also continued to interview students, review faculty publications, and dig deep into faculty lifestyles.[73] He insisted that evolution was still taught at Baylor, in spite of Dow's resignation and the avowals of orthodoxy by President Brooks.[74] Norris also learned that one member of the faculty had left his wife and child and that this professor was being sued for divorce. Some faculty resigned under the scrutiny and pressure, but most agreed to sign a letter to the board of trustees affirming their orthodoxy.[75] This finally brought Norris to "Great Rejoicing!" He claimed a great victory for fundamentalism, as the faculty agreed to several key tenets, including such fundamentalist standards as the Genesis account of creation, the "supernatural elements in the Christian religion," the inerrancy of the Bible, and the virgin birth, deity, and miracles of Christ.[76]

Although Norris's attention was soon diverted by the Scopes trial, he continued his attacks on Baylor's President Brooks in a desultory manner for years. He deluged Brooks with piles of accusatory and time-consuming correspondence throughout the decade. The stress of constant scrutiny led Brooks to complain privately in late 1928 that Norris had "sought to damage Baylor University for several years" and warned a correspondent to ignore "anything he says."[77]

Norris's long-term dedication to college activism was typical of fundamentalists during the 1920s. Further, in spite of the attention paid—at the time and by historians—to the fundamentalists' efforts to ban evolution from high schools in the 1920s, it was actually fundamentalists' concerns about the teaching at colleges and universities that first sparked fundamentalist interest in educational activism. In the field of higher education, that activism was as diverse as the fundamentalist movement itself. Many fundamentalists tangled with the leadership of their state universities. They fought not only to ban evolution but to reassert traditional Protestant intellectual hegemony. To do so, fundamentalists exerted pressure on administrators in every way they could, including publicizing the teaching of evolution and atheism and attempting to pass legislation banning such teaching from public higher education. Although these efforts failed to accomplish the goal of banning such teaching from any major public research universities, they did prompt professors to view their own work with an eye to the potential firestorm that fundamentalists might raise about controversial ideas. For some, as Chapter 7 will describe, this meant self-censorship to avoid the blistering attacks of fundamentalist activists.

Many public university administrators also reluctantly abandoned their traditional dual role as religious and educational leaders. Like President Birge at the University of Wisconsin, they were forced to defend the values of academic freedom and open academic inquiry, in spite of their own devout Protestant beliefs. Fundamentalists' battles with denominational schools also had varied results. At some schools, such as the University of Chicago, fundamentalist opposition merely prompted a more explicit identification with theological modernism and intellectual liberalism. At other schools, such as Baylor, fundamentalist pressure prompted faculty to adopt a fundamentalist creed. These first years of controversy during the 1920s established battle lines at colleges and universities across the nation. Overall, as historian George Marsden has demonstrated, fundamentalists did not win the battle for control of American colleges and universities. By the time of the Scopes trial in 1925, leaders such as Birge and Poteat were solidly in the mainstream of American higher education, rather than fundamentalists such as Norris, Bryan, and Riley.[78] Fundamentalists learned to their dismay of their severely diminished clout in mainstream academic culture. The early campaigns of those activists, however, opened the first front in the twentieth century culture wars. Many fundamentalists began to consider themselves a victimized minority group, and college campuses became the intellectual battleground on which they fought for their minority rights. Students, faculty, parents, and administrators became keenly aware of the promise and peril of fundamentalism. Fundamentalist students at nonfundamentalist colleges had been warned time and again about the "atheistic influences" to which they might be subjected at a mainstream denominational college or public university.[79] Faculty had been put on notice that their publications, lectures, lifestyles, and even private conversations would be scrutinized by an aggressive and hostile fundamentalist community. Administrators added a new danger to their list of political quagmires. One misstep and they could find themselves in a protracted struggle, sometimes for the very existence of their school. Parents were given another concern to add to their worries. Liberal parents would have to fret about fundamentalist pressure on their children's college, while fundamentalist parents worried that their son or daughter might come home "an avowed infidel."[80]

CHAPTER 4

Early Legislative Battles

In the spring of 1922, Kentucky's state legislature became embroiled in a bitter battle over the nation's first statewide fundamentalist school bill. Frank McVey, president of the University of Kentucky, fiercely opposed the measure.[1] During the heated debate, State Senator Harry F. Greene taunted McVey with the challenge, "If he is not teaching evolution what is he hollering for? If the university is not teaching evolution this bill does not hit it."[2] In fact, Kentucky's so-called antievolution bill would have had a much wider scope. The bill banned evolution, but it would also have prohibited teachers in Kentucky's public schools from teaching any idea that might challenge students' religious beliefs. As in Kentucky, state legislators nationwide took sides in fundamentalist school campaigns during the 1920s. Like Senator Greene, most lawmakers considered the central issue to be the teaching of evolution, even as they debated bills that often made much broader claims.

Fundamentalists led the drive for such "antievolution" bills, but to achieve even a hope of success fundamentalists had to reach out to nonfundamentalist conservatives. They articulated arguments that appealed to popular opinion beyond the tenets of fundamentalist theology. In most cases, they argued that the teaching of evolution represented a sort of stalking horse for a cluster of distressing trends in modern culture. Its elimination from public schools, many fundamentalist activists argued, would exclude a range of antipatriotic, anti-Christian, antitraditional ideas from those schools. In the years before the Scopes "monkey" trial, these legislative efforts did not succeed as well as many fundamentalists hoped, but many contemporary observers concluded they were the rising tide in American educational policy.[3] Furthermore, even when fundamentalist school laws failed passage, they demonstrated the beliefs and desires of fundamentalist supporters. As many contemporary observers noted, this campaign hoped not only to ban a single scientific theory but to maintain public schools that would defend and promote the presumed evangelical Protestant faith of their students.

Fundamentalist leaders never tried to disguise their ambitious goals. They talked openly and often about their grandiose hopes for reclaiming cultural and religious control of public schooling. Fundamentalists believed that public schools had a responsibility to combat atheism actively. William Jennings Bryan, for instance, described the goal of the antievolution movement as keeping "the religion of the school children protected."[4] Bryan insisted that "Atheists, Agnostics, and Higher Critics begin with Evolution: they build on that."[5] According to Bryan, evolution itself and its attendant materialistic teachings led to atheism. It was only as such that it formed a "menace to fundamental morality." The danger of teaching evolution as fact, Bryan believed, was not merely that students might learn about Darwin's theory. As did many leading fundamentalists, Bryan supported such teaching if done carefully. The danger came when such teaching led students away from their Christian faith. Not only fundamentalists ought to worry about such declension, in Bryan's opinion. Society as a whole could not survive the results of such teaching. Bryan grimly predicted that "all the virtues that rest upon the religious ties between God and man" would soon collapse as a result of widespread atheism if such teaching did not stop.[6]

Antievolution activist T. T. Martin explained how the teaching of evolution could cause this vicious downward spiral. He cited an elementary textbook: Harold W. Fairbanks's *Home Geography for Primary Grades*. In this classroom text, Fairbanks described the evolutionary history of sea mammals such as seals and whales. According to Fairbanks, "Their grandfathers lived upon the land ever so long ago. . . . They used to go into the water for food and at last they spent the most of their time there. Their bodies and legs became changed so that they could swim or paddle through the water."[7] According to Martin, young Christians reading this simple evolutionary tale would soon conclude that the Genesis story of creation could not also be true. In that story, God created humans and animals in roughly their present state. As a result, in Martin's telling, students must soon conclude, "Listen to those lies in the Bible!" If, as Martin asserted, Jesus vouched for the veracity of the Scriptures, then He must also become suspect in children's minds. Soon, inexorably, such teaching would turn a young child to atheism, and, according to Martin, "That child's faith in the Saviour is gone forever, and her soul is doomed for Hell; and with your taxes, you paid to have it done." Public schools, in this logic, must prevent atheism; banning evolution was merely the single most obvious place to begin.[8]

For Martin as for Bryan, that atheism and resulting damnation were the true dangers to be prevented by maintaining fundamentalist control of public schooling. Evolution was only one manifestation of what fundamentalist educator Alfred Fairhurst attacked as the many-headed "godless teaching" that went on in colleges and universities. Such teaching served only for the "propagation of atheism," Fairhurst warned.[9] Other leading fundamentalists agreed. William Bell Riley described what he called the "Evolutionary Controversy" as nothing less than an all-out "war" that would not end until public schools no longer harbored "teachers and textbooks that scorn the Christian faith."[10] As it did for other leading fundamentalists, evolution for Riley served as a bellwether for any teaching that might threaten Christian belief. Riley and other leaders often used "evolution" as shorthand for any atheistic, skeptical, or materialistic teaching that encouraged students to question the tenets of their faith.

Not only the national leaders of the fundamentalist movement espoused this understanding and use of the term evolution as a stand-in for any number of ideas considered pernicious. Rank-and-file fundamentalists often agreed. One layman from Alcolu, South Carolina, put it simply: "The doctrine of evolution," he explained, "is the spiritual path that leads to Sodom." Sin and damnation resulted from any denial of the basic truths of Christian faith. Evolution, in this fundamentalist's understanding, was merely one particularly underhanded way to encourage students to deny those truths, in the guise of scientific theory.[11] One Indianapolis women's group described their fundamentalist school campaign as a fight for nothing less than "the preservation of the public schools of America, the Bible and the faith of our fore-parents." This group, like other fundamentalists, agreed that evolution was both part of this wide-ranging campaign and a symbol of everything it fought against.[12]

The wider fight to ensure that public schools supported broadly Protestant teachings often manifested itself in state-level laws that both supporters and opponents labeled simply "antievolution." In Texas, for instance, State Representative J. T. Stroder backed a series of bills to ban evolutionary teaching from Texas schools. But Stroder's campaign targeted more than just evolution. To support his charges of atheistic teaching, Stroder circulated a speech given by Herbert L. Willett of the University of Chicago. Stroder reported to his constituents that Willett's speech repudiated fifteen fundamental doctrines of Christianity. Such doctrines, including but not limited to evolutionary teaching, formed the real target of Stroder's ire. Stroder challenged his constituents to join him in the fight. "Parents!" he demanded in a form letter, "How many of you are aiming to sit idly by and not make a strong protest to oust this blasphemous demon from all our schools?" In this case, the "blasphemous demon" from which students needed protection espoused not only evolutionary theory but fourteen other materialistic ideas besides.[13]

The debate in Tennessee's state Senate over its successful antievolution bill demonstrated the ways supporters viewed the real threat from evolutionary teaching. The bill's supporters believed such teaching could lead to a host of frightening results. The bill's Senate sponsor, John A. Shelton, outlined the real meaning of his drive to safeguard schools. "Whenever the belief in the existence of a human soul is destroyed," argued Shelton, "you will have destroyed the happiness and hope of mankind." As did national leaders of the fundamentalist movement, Shelton identified his goal as protecting the evangelical Christian faith of Tennessee's public schoolchildren. Shelton did not merely want to ban one idea that might threaten that faith. Rather, he argued that the primary role of public schooling must be to safeguard that belief. L. D. Hill, then speaker of the state Senate, agreed that the bill must become law in order to maintain America's status as a "Christian nation." He argued along lines similar to those of fundamentalists leaders and rank-and-file from across the nation. If children were taught evolution, Hill argued, "they [would] never believe the Bible story of the divine creation." State Senator Whitfield similarly insisted that "nothing contrary to [Christian] belief be taught" in Tennessee's public schools. Whitfield lamented that in his county students were "being taught in the public schools to be infidels, that there is no immortality of the soul, that there is no heaven nor hell." Such ideas, according to the senator, could not be allowed in public schools. That, Whitfield explained, was the reason for his support of the so-called antievolution bill. For state Senators Hill, Stroder, and Whitfield, and for other fundamentalists, antievolution legislation meant maintaining public schools in which Christian belief would be protected and encouraged.[14]

Before Tennessee's Senate debated its Butler Bill in 1925, many other state legislatures held similar debates. Kentucky was the first state to consider some sort of fundamentalist-inspired school legislation in early 1922.[15] South Carolina followed in the same year.[16] In 1923, several states considered similar legislation. In Georgia, representatives introduced two resolutions, one of which received a favorable report from the legislative committee on education.[17] In Texas, an ambitious antievolution bill sailed through the House of Representatives with a 71 to 33 majority. The bill received a favorable recommendation from the Senate's educational committee. Nevertheless, the bill stalled on the Senate calendar.[18] Later that same year, a strongly worded resolution passed the Texas House by an overwhelming margin of 81 to 9. However, the resolution stalled in the state Senate.[19] Also in 1923, the state legislatures of West Virginia, Alabama, Iowa, and Tennessee all defeated some form of fundamentalist school law.[20] Meanwhile, Florida's state legislators approved a nonbinding resolution condemning any teaching that led students away from traditional Protestant belief.[21] Soon after, Oklahoma approved a law restricting state-supplied textbooks to those without neo-Darwinian evolutionary theory.[22] The next year, California's State Board of Education appointed a panel of nine college and university presidents to consider a restriction on state textbooks. The panel found no problems with the textbooks in question, yet the state board of education issued a warning to school teachers to restrict their teaching of evolution nevertheless.[23] North Carolina's Board of Education similarly voted to ban a biology book, which they claimed contradicted the creation story in Genesis.[24] The last major fundamentalist-driven legislative action of 1924 took place in Washington, DC, as the United States Congress quietly passed an amendment prohibiting DC teachers from teaching any idea suggesting "disrespect of the Holy Bible."[25] In 1925, several states considered similar legislation. In West Virginia,[26] North Carolina,[27] Georgia,[28] and Florida,[29] the bills were defeated. In Texas, although the bill was defeated, the governor ordered publishers to edit out all mention of neo-Darwinian evolution from textbooks sold in Texas.[30] In Tennessee, the Butler Bill finally became law on March 23, 1925. It targeted the teaching of evolution relatively narrowly, compared to the broad language of several other state bills. Even this law, however, preserved a special role in public schools for Christian belief. The law stipulated not only that evolution be banned but that no theory be allowed to challenge students' belief in the creation story as told in Genesis.[31]

Some of these state legislatures borrowed the sweeping language of Kentucky's 1922 House Bill 191. In 1923, bills considered in Alabama and Tennessee, as well as the resolution that successfully passed in Florida, would have banned the teaching of "atheism, agnosticism, Darwinism, or any other hypothesis that links man in blood relationship to any other form of life."[32] The resolutions considered in both Texas and Georgia included similarly broad requirements.[33] Some state legislatures considered much more sweeping fundamentalist-backed educational laws. In Iowa, the House considered a bill that would introduce a curriculum of religious education into public schools.[34] Iowa fundamentalists believed the measure would serve to banish evolution from schools as well as to introduce a measure of nonsectarian Protestant doctrine. They rallied support from the Women's Christian Temperance Union and Iowa Lutherans for their broadly worded measure "for the imparting of moral and religious instruction to pupils in the public schools."[35] In Tennessee, a bill to "make it unlawful to employ atheists as teachers in Public Schools of the State" enjoyed considerable support. The bill emerged from the committee on education with a positive recommendation, and was only defeated by one vote.[36]

Of the twenty-five bills, amendments, and resolutions considered among state legislatures and by the U.S. Congress in the years preceding the Scopes trial, most included some wider affirmation of traditional Protestant doctrine. Only nine of these early bills limited themselves strictly to banning evolution. Sixteen explicitly preserved a special role for the Bible or Protestant religiosity.[37]

Due to the widespread popularity of traditional Protestant school policy among conservatives, fundamentalists were able to rally large sectors of support from among nonfundamentalist ranks. Many conservative Protestants who wanted a stronger role for traditional Protestant doctrine in schools did not consider themselves fundamentalists. For example, conservative Lutherans often stood on the sidelines of the fundamentalist movement in the 1920s. Although many conservative Lutherans "rejoiced" in the successes of fundamentalists, they remained aloof from the early fundamentalist movement as a whole. Nevertheless, Lutherans occasionally supported fundamentalist-backed school legislation.[38]

Other conservative Christians outside of the fundamentalist movement also opposed the increasing secularization of public schools. Catholics were the biggest such group. Many Catholics opposed the teaching of evolution and favored some infusion of religion into secular schools but feared Protestant attempts to control the types of devotions allowed.[39] Besides, conservative Catholics had an extensive network of Catholic parochial schools to provide their children with a Darwin-free education.[40]

Some Catholics, however, directly supported fundamentalist efforts to safeguard traditional religious practices in public schools. Many Catholics agreed that evolutionary theory was not the only threat, but rather one outstanding manifestation of a much more sweeping change in intellectual culture that would drastically undermine what the Reverend William Hornsby considered "the foundations of the Christian religion." Hornsby, a Jesuit professor at St. Mary of the Lake Seminary in Mundelein, Illinois, expressed "sympathy" with the fundamentalist movement's goals, especially including the goal of banning the teaching of evolution from public schools.[41]

As we have seen in Chapter 2, some conservative Catholics even identified themselves as fundamentalists. Far more common, however, were Catholics such as the Reverend Hornsby. Like Hornsby, Patrick Henry Callahan, former leader of the Knights of Columbus, insisted that "there is a vast amount of sympathy for Mr. Bryan and the State of Tennessee among the Catholics of America." As a Christian, Callahan fumed at the "ribald abuse" that had been directed at the movement to maintain Christian religiosity in public schools.[42] Similarly, Michael J. Lavelle, a Catholic priest in New York City, insisted that public schools must include "religious instruction." Like many Catholics,

Lavelle worried that such instruction in the past tended to "offend or transgress the rights" of religious minorities. Nevertheless, without "school Bible work," Lavelle worried that public school graduates might not become "faithful servants of their God."[43]

Fundamentalists and conservative Catholics often agreed that evolutionary theory represented the cutting edge of a new wave of materialistic thought. Catholic writers such as George Barry O'Toole and Louis T. More blasted the intellectual credibility of organic evolution. In spite of More's and O'Toole's Catholic beliefs, which most Protestant fundamentalists considered anathema, O'Toole's *The Case against Evolution* (1925) and More's *The Dogma of Evolution* (1925) became fundamentalist favorites as fundamentalists pushed for school legislation.[44]

Alfred McCann's *God—Or Gorilla* (1922) became another standard of the fundamentalist canon. The spirited prose of the Catholic writer hardly took a back seat to Protestant polemics. McCann damned "the materialistic evolutionists, falsifying their unscientific deductions and misrepresenting the honest research of the laboratories." McCann warned readers of the widespread influence of such dangerous evolutionists. "There is left scarcely a channel of public information," McCann complained, "through which does not flow the false conviction that man's origin as a descendent of the ape has been 'scientifically demonstrated.'"[45] As did other prominent Catholic intellectuals, McCann maintained personal contacts with leading fundamentalists as fundamentalists promoted new school laws.[46]

As did conservative Protestants, conservative Catholics worried that the teaching of evolution and other doctrines undermined the credibility of their theology. More secular conservatives also found the teaching of evolution objectionable. Fundamentalist leaders often appealed to more secular conservatives by arguing that secularism and evolution led to social chaos. For example, Bryan claimed that evolutionary theory "plunged the world into the worst of wars, and is dividing society into classes that fight each other on a brute basis."[47] An American public still reeling from the brutality of trench warfare in France and labor and race conflicts from Seattle to Chicago needed few reminders of the ways modernity could result in horrifying ends.[48] Other nonfundamentalist conservatives worried that an educational system that had "lost its element of spiritual truth" would raise a weak generation unable to defend America against foreign enemies.[49]

Many of these nonfundamentalists took part in the campaign to ban evolutionary teaching and other liberal or secular ideas from public schools. As political scientist Michael Lienesch has recently argued, fundamentalists made themselves particularly effective at using the evolution issue to attract support from beyond the ranks of self-identified fundamentalists.[50] Fundamentalists published reams of antievolution literature in every format imaginable. Such literature provided fence-sitters and committed activists alike with readymade arguments in favor of protecting a favored role for Protestant doctrine in public schools. Fundamentalist activists also traveled widely in the years before the Scopes trial, debating evolutionists and making speeches before state legislatures and other civic bodies. In some cases, fundamentalists took a much more active role, writing the bills that lawmakers proposed. Once the bills had been introduced, itinerant fundamentalists pushed local congregations of several denominations to send supportive letters and petitions to state lawmakers. In addition, fundamentalists organized local and national groups to coordinate legislative efforts. In some cases, the effect of fundamentalist efforts was decisive. In other places, fundamentalist activism played a more indirect role. As the "antievolution" movement among state legislatures and state boards of education gained steam in the years before the Scopes trial, energetic fundamentalists wielded enormous influence over the campaign.

One of the most potent tools wielded by fundamentalists was the press. Independent fundamentalists cranked out inflammatory volumes that sold thousands of copies. Perhaps the best known of these was Martin's *Hell and the High School* (1923). Once the Scopes trial became a media event, Martin grabbed the attention of visiting journalists in Dayton, Tennessee with his nightly sermons and prominent newsstand. The newsstand boldly advertised his provocatively titled work, attracting the attention of partisans on all sides. Martin dubbed himself the "Blue Mountain Evangelist," and by the time of the Scopes trial he was one of the nationally recognized leaders of the fundamentalist movement. Martin challenged readers of all beliefs to take up the antievolution crusade. "Ramming poison down the throats of our children is nothing," Martin accused, "compared with damning their souls with the teaching of Evolution." Fundamentalist parents might shudder at the threat to their children's spiritual well-being. But Martin also used arguments that would appeal to nonfundamentalists. Martin mixed appeals to fundamentalist theology with manipulations of gender stereotypes and patriotic history. He asked readers, "Where is your Christian manhood? Where is the spirit of those who came over in the Mayflower? Where is the spirit of 1776?"[51]

Like Americans for every political cause, Martin and other fundamentalist activists printed thousands of copies of pamphlets to address more specific issues. Martin himself published an address he had given in Los Angeles in

1923 as citizens in California considered legislative action. In the pamphlet, Martin again used every rhetorical weapon at hand, not only invocations of fundamentalist theology. For instance, he appealed to citizens in the emotion-laden historical memory of the Alamo: "One man slunk out [of the Alamo]; WILL YOU? Drive out every member of the local board of trustees who will not do his duty in driving out every Evolution teacher and book from all of our tax-supported schools, from primary to university. Drive out every legislator who will not go to the limit. . . . If we don't, our children are doomed."[52] Other prominent fundamentalists also produced pamphlets to address issues in specific states. In North Carolina, fundamentalist pastor S. J. Betts printed pamphlets to rebut a fundamentalist-bashing speech by the president of Wake Forest College. "Some would want we [*sic*] fundamentalists to keep quiet while this destructive propaganda continues its work by evolution teachers," Betts argued. "We have already kept quiet too long."[53]

Fundamentalists also controlled larger presses. The Moody Bible Institute of Chicago (MBI) printed and distributed hundreds of thousands of copies of pamphlets and tracts through its Bible Institute Colportage Association (BICA). Some of the titles reproduced speeches by well-known antievolutionists, such as Bryan's *The Bible and Its Enemies*. Although such influence is difficult to trace, the impact of the Chicago press is evident. In North Carolina, for instance, superior court judge Thomas J. Shaw avidly read the BICA literature. According to Shaw, the BICA-published *Modern Education at the Cross-Roads*, by M. H. Duncan, was one of the biggest influences on his antievolution activism.[54] The Moody Bible Institute also reached thousands of grassroots activists through its monthly magazine, the *Moody Bible Institute Monthly*. With around twenty thousand subscriptions in 1921, the *Monthly* had a wide influence. By 1923, the magazine claimed readers in forty-seven states and from at least fifty-one denominations. The *Chicago Tribune* and other secular newspapers often quoted the *Monthly* as the voice of the fundamentalist movement.[55]

Throughout the 1920s, the *Moody Bible Institute Monthly* published important articles about public schools and the legislation drive, essential reading for many fundamentalist school activists. For instance, in February, March, and April 1921, the *Monthly* published a lengthy three-part series by Seventh-day Adventist geologist George McCready Price, "Modern Problems in Science and Religion." In this series, Price laid out his theory of "flood geology," which at the time only a small fraction of fundamentalists believed. According to Price, the only truly biblical position was that of a young earth, devastated by a literal worldwide flood at the time of Noah. Although this argument did not convince the majority of fundamentalists at the time, 1920s-era fundamentalists still looked to Price as their leading scientific representative, and this series in the *Monthly* introduced many activists to Price's ideas.[56]

The *Monthly* also ran less intellectually intense antievolution items. Poems from contributors across the nation poked fun at the evolutionists. A Michigan fundamentalist wrote to the Chicago office with her attempt to share a laugh at the expense of such a ridiculous theory:

> Why Not Be Up-to-date?
> If yer sprung from apes an monkeys
> In an evoluti'n way,
> Will yer jest perlitely tell me
> Why did evolution stay
> Way back thar in early ages,
> Sted o' keepin' up-to-date?
> Why not man a present product
> Of the monkey and his mate?

Poems and other light pieces provided readers with more than convincing antievolution arguments. Any of the tens of thousands of readers of the *Moody Bible Institute Monthly* were provided with both intellectual ammunition against evolution and a sense that they were part of an international community of fellow believers. As the fundamentalist movement developed in the early years of the 1920s, this antievolution literature broadened support for fundamentalists among the wider conservative public. It also served to convince fundamentalist readers that opposition to evolution was a key element of their inchoate desire to regain broader control over public schooling.[57]

Bryan was another prolific publisher. He eagerly sought publishers for his major antievolution speeches, such as *The Bible and Its Enemies* (1921), and *The Menace of Darwinism* (1919?). Bryan sent hundreds of copies of the latter pamphlet to influential fundamentalists and nonfundamentalist educators. In it, he laid out many of the common arguments of his antievolution campaign. Evolution, he said, led children away from true biblical morality. If unchecked, such teaching produced generations of militaristic, amoral youth, as had happened in Germany. Tax-paying parents also had a right to control the education of their children. If a majority objected to any teaching, then teachers should comply. Most importantly, the Darwinian theory of natural selection was not true.

It was only a "hypothesis" and "not only groundless, but absurd and harmful to society."[58]

Some educators had not thought deeply about the meaning of evolution until they received signed copies of this pamphlet directly from Bryan. The president of the University of Florida and the principal of the state teacher-training college of Pennsylvania were both flattered by the attention from such a prominent public official.[59] The same must have been true for some of the Florida ministers to whom Bryan sent copies, with his compliments. Whether or not they shared his views, Protestant ministers in the state and many prominent Catholics also received free copies. Using this broadcast approach, Bryan built a base of support among both committed fundamentalists and conservative allies. Bryan's publishing efforts provided recipients with a convincing, articulate, and memorable educational argument.[60]

It did not take long for the controversy to capture the interest of the mainstream press. In most cases, journalists assumed with some justification that the only target of fundamentalist school activism was the teaching of evolution. Journalists helped create a perception that the much wider goals of fundamentalist school legislation could be summed up as "antievolution." In early 1922, for instance, the *New York Times* invited Bryan to present his side of the evolution argument. Then the *New York Times* offered similar space to prominent proevolution writers. Bryan seized his chance to rehash the arguments he had made in his favorite antievolution speeches.[61] The exposure in the *New York Times* allowed Bryan to get his arguments across to a wider audience and to show that antievolutionism had a solid intellectual footing. Even before this publicity, however, Bryan had been reaching plenty of newspaper readers with his syndicated "Weekly Bible Talks." By early 1923, an estimated ten to twelve million Americans read these columns in approximately one hundred newspapers nationwide.[62] In some states, these syndicated columns had an important effect on the antievolution crusade. In North Carolina, for instance, the column ran in three of the state's most widely read newspapers. Fundamentalist and nonfundamentalist readers alike could read Bryan's weekly arguments about the propriety of banning evolutionary theory from any tax-supported classroom.[63]

More than for their publishing efforts, however, fundamentalists attracted notice in the years before the Scopes trial for their personal lobbying for state antievolution laws. The first state to consider such legislation was Kentucky, and fundamentalists gave the state their full attention. William Bell Riley saw the Kentucky legislation as a perfect chance to put the new World's Christian

Fundamentals Association (founded in 1919) to work. Riley set up twenty-two meetings around the state in the spring of 1921 and led a team of fundamentalist stump speakers.[64] Bryan also toured, even giving an antievolution speech to a joint session of the state legislature.[65] In spite of this activism by national leaders, local fundamentalist pastor John W. Porter took the lead in Kentucky. Porter pressured the Baptist State Board of Missions to form a committee to lobby for a state evolution bill. Porter also published his views in editorials around the state and in his 1922 book *Evolution—A Menace.*[66]

Porter supplied the wording in the antievolution bill proposed in the state House of Representatives in January 1922. This wording demonstrates the much wider goals of the so-called antievolution movement. The Kentucky bill would have outlawed not only the teaching of evolution but also the teaching of "Darwinism, Atheism, Agnosticism, or the Theory of Evolution."[67] A state Senate amendment made this wider implication even more explicit. The amended bill would have prohibited any public library from owning any materials "containing such teaching that will directly or indirectly attack or assail or seek to undermine or weaken or destroy the religious beliefs and convictions of the children of Kentucky."[68] Had it passed, this bill would have had much wider potential impact than simply banning evolution. What might it have meant to be accused of teaching "agnosticism"? What books might have been banned on suspicion that they might "indirectly" challenge students' faith? These proposed bills demonstrated the grandiose educational hopes of fundamentalists in the early years of the 1920s. Not only did they hope to free schools of evolutionary teaching, but they hoped to legally mandate a traditional educational system dedicated to protecting and even fostering Protestant faith in its students. Nor was this ambitious legislation defeated soundly. The House Committee on Rules reported the bill favorably and it only lost by one vote. The author of the House bill, George Ellis, noted privately that legislative horse-trading had allowed him to accept the bill's narrow defeat in return for a promise from his fellow lawmakers to eliminate informally all atheistic, agnostic, or evolutionary teaching from Kentucky's public schools.[69]

The Kentucky fight became a model for many other state legislatures. Press coverage brought the issue to the attention of many lawmakers. Supporters of evolution noted with alarm the widespread popular support for the legislation in Kentucky, rightly sensing the fundamentalist school movement was rising elsewhere.[70] Both evolution supporters and opponents were impressed with the way fundamentalists pulled together national speakers and grassroots organizers. Even in defeat, fundamentalists took heart at the prospect of similar struggles in other states. In some cases, the Kentucky fight influenced others immediately. G. W. Moothart, state representative to the Oklahoma general assembly, soon planned to copy the Kentucky bill. He hoped to introduce legislation to prevent "not only our high schools but all State Educational institutions in any way supported by State taxes, to have any teachers, or use any textbooks, which present Darwinian Evolution, or any other Author's works of kindred nature." For the details, Moothart asked Bryan for a copy of the Kentucky bill.[71] In the end, the Oklahoma state legislature, with the strong support of the state Ku Klux Klan organization, substituted a unique textbook bill. It promised free textbooks to public school students so long as the textbooks did not mention evolution.[72]

Bryan and other leading fundamentalist activists toured indefatigably in support of state bills to ban evolution. The West Virginia legislature considered a much more specific antievolution bill in early 1923. Unlike Kentucky's sweeping claims, the bill considered by the House of Delegates would only have banned "the teaching of the Darwin theory of evolution in the public schools of West Virginia."[73] Nevertheless, in his speech to a joint session of the state legislature, Bryan made sure legislators knew of the wider religious ramifications of teaching evolution. "Evolutionists," Bryan thundered, "rob the Savior of the glory of virgin birth, the majesty of His deity and the triumph of His resurrection. They weaken faith in the Bible by discarding miracles and the supernatural and by eliminating from the Bible all that conflicts with their theories. They render that book a scrap of paper." After his speech to the legislature, Bryan traveled across West Virginia, reminding voters of similar points.[74] Throughout the spring and summer of 1923, Bryan made similar tours and speeches to legislatures in Georgia, Texas, and Florida and turned down an invitation to address the state legislature in Iowa.[75]

Across the nation, fundamentalists mobilized to support these school bills and resolutions. In Iowa, local fundamentalists recruited allies from moderate Protestants and conservative Lutherans to introduce broad legislation that included the prohibition of evolution and of other teachings perceived as "anti-Christian." As did many early fundamentalists, they felt confident that their transformative school bill would be a rapid success. Iowa activists, like fundamentalists elsewhere, expressed shock and dismay when their bills met determined and successful opposition.[76]

Texas, for its part, was home to prominent Baptist fundamentalist J. Frank Norris, who embraced the legislative campaign with all of his considerable energy. Norris spoke to a joint session of the Texas legislature in support of the 1923 bill to ban evolution.[77] Norris warned lawmakers; if they "let atheistic evolution have its way . . . how long will it be until the recent incident in Russia is repeated in the United States, where professors met in the streets of Moscow and burned God in effigy!"[78] Norris also brought the annual conference of the World's Christian Fundamentals Association to Fort Worth in 1923. In Norris's opinion, lawmakers would certainly feel pressured by the proximity of so many dedicated fundamentalist pastors and activists.[79] WCFA founder William Bell Riley also toured Texas, whipping up crowds with his impassioned attack on the deviltry of evolution; and although Bryan was unable to make a trip to Texas, he privately urged Texas governor Pat Neff to accept antievolution legislation. Bryan also wrote to sympathetic Texas state legislators to assist them in their formulation of a bill.[80]

Although the Texas bill was narrowly defeated, Bryan had more success in his adopted home state of Florida. While Norris and Riley hectored Texas crowds and legislators with reasons to fear evolution, Bryan applied more direct pressure to Florida lawmakers. He asked sympathetic legislators to distribute copies of his antievolution speeches to the entire state assembly. Bryan also offered nuts-and-bolts support to state legislators, by sending them copies of earlier bills, including the Oklahoma bill that offered evolution-free textbooks.[81] Meanwhile, Bryan also courted grassroots support. His publisher, Fleming Revell, agreed to send copies of Bryan's *The Menace of Darwinism* to every Protestant minister in the state.[82] Bryan's strategy worked. The Florida state legislature agreed in 1923 to yet another broadly worded resolution. They condemned the practice of "teach[ing] or permit[ting] to be taught atheism, agnosticism, [or] Darwinism." The resolution, as Bryan had suggested, carried no penalty provision. Florida's lawmakers agreed with Bryan that their resolution should make a symbolic statement of the state's antievolution stance, rather than become a source of criminal prosecution.[83]

All of this public exposure of the evolution question helped to fuel the intense interest in the Scopes trial in the summer of 1925. Even before Tennessee had passed its controversial antievolution law, fundamentalist activists applied the same kind of pressure in Tennessee that they had been using in other states. In the words of one dismayed local journalist, "gadfly evangelists swarm[ed] through the state."[84] Just as they had elsewhere, fundamentalists toured through Tennessee as early as 1923, delivering polished polemics. Riley traveled throughout the state in 1923.[85]

Bryan also spoke in Tennessee in early 1924, before the Butler antievolution bill had been introduced. As usual, Bryan's combination of intellectual argument, conservative Protestant theology, down-home common sense, and emotional appeal found its target. One member of Bryan's Nashville audience that January evening was local attorney W. B. Marr, who was so moved by Bryan's talk that he had five hundred copies of the speech printed and delivered to state legislators and other influential Tennesseans. W. J. Murray, a Primitive Baptist preacher in Nashville, was similarly impressed by Bryan's arguments. Murray incorporated some of them into his next sermon, which had in turn a powerful impact on one of the members of the congregation. That congregant, John W. Butler, soon introduced the antievolution bill into the state House of Representatives.[86] The Butler Act, as Tennessee's antievolution law became known, was the first in the nation to ban the teaching of evolution in public schools. Unlike many earlier laws, it did not explicitly ban the teaching of atheism or agnosticism, but it did preserve a special role for biblical Christianity. According to the Tennessee law, teachers and schools must not teach "any theory that denies the Story of the Divine Creation of man as taught in the Bible."[87]

But such indirect influence, while important, did not form the extent of fundamentalist activity in the passage of the Tennessee law. Bryan also corresponded with John Shelton, the author of the state Senate's version of the bill. As he had with other bills, Bryan pressed for a bill like Florida's, without any criminal penalty attached. If Shelton had heeded Bryan's advice, then the Scopes trial never would have been possible.[88]

Even before the headline-grabbing events of the trial, however, the bitter controversies stirred up by fundamentalist educational activism had begun to exert pressure on the image of the relatively new fundamentalist movement. Early in 1922, for instance, the superintendent of Cleveland's public schools told the attendees of the annual meeting of the National Education Association that the antievolution movement would soon be "completely discredited" due to the absurd tactics of fundamentalists. Other speakers called the movement intellectually childish, and, like many other foes of fundamentalism, accused Bryan of being a "medievalist."[89] George Bernard Shaw called Bryan's attacks on evolution in schools "the stigmata of blockhead" and asserted, "What [Bryan] calls fundamentalism I call infantilism."[90] Fundamentalists came under similarly bruising attack from fellow evangelical Protestants as well. One liberal Protestant objected to the "Heresy-Hunting" at denominational schools.[91] An evangelical defender of liberal denominational colleges blasted Riley and other fundamentalist educational activists as nothing but a gang of "reactionaries."[92] Other liberals attacked the fundamentalist movement as a bastion of "doctrinal obscurantists,"[93] and many agreed that fundamentalist educational policies represented nothing less than "a deliberate reversion to ignorance."[94]

Until the summer of 1925, however, fundamentalists could take comfort from the wide support enjoyed by their movement on all fronts. The new movement attracted adherents of all kinds, and achieved remarkable successes. As long as it did so, fundamentalists could dismiss the rancorous attacks. The unprecedented publicity surrounding the Scopes trial, however, tipped this struggle in the liberals' favor. Despite the earnest efforts of many fundamentalists, a popular stereotype of fundamentalism soon became its effective image. Instead of a nationwide movement of concerned Protestant activists in the best tradition of American evangelicalism, fundamentalism soon became tarred as a cultural vestige, functional only in remote Southern pockets of unenlightened intellectual backwaters.

CHAPTER 5

Of Monkeys and Men

Kelso Rice, the court policeman in Dayton, Tennessee, struggled to make his voice heard over the laughter and applause in the loud, crowded courtroom. "People, this is no circus," he demanded. "There are no monkeys up here. This is a lawsuit, let us have order."[1] The roomful of East Tennesseans, newspaper reporters, big-city lawyers, and curious onlookers could have been forgiven for assuming they were at a circus. For days, the East Tennessee town of Dayton had shown all the signs of it. The streets were full of buskers, snack stands, and carnival games. The hot July sun roasted the courthouse until even this latest "Trial of the Century" was forced outdoors under the shade of some cottonwood trees. For the first and only time in its history, Dayton was the most talked-about small town in the world. Reporters telegraphed hundreds of thousands of words daily from the town to the waiting world, as America held its breath for the outcome.[2]

For many Americans at the time, the Scopes trial typified the battle over the teaching of evolution. But even this archetypical evolution battle hinged on much more than simply the teaching of evolution. The defense strategy relied on proving that the Tennessee antievolution statute mandated an unconstitutional assertion of fundamentalist theology into public education. The defense sought to prove that evolution only represented a cultural and theological threat to fundamentalist interpretations of Protestantism. The Butler Act stated that no teacher may teach evolution or "any theory which denies the story of divine creation of man as taught in the Bible." The defense team at the famous trial argued that this definition only defended one kind of Christianity, the fundamentalist sort. Other Christians, indeed most leading Protestant theologians, could testify that a conflict between the creation story in the King James Version of Genesis and that of evolution did not constitute a denial of Christianity altogether. In other words, the defense strategy rested on the assertion that Tennessee's schools, even if they taught evolution, still would not threaten the Christian faith of public school students. In the event, Judge Raulston ruled that such defense evidence was inadmissible. John Scopes was convicted on the narrow question of whether or not he had taught evolutionary theory to his students.[3]

The world-famous representatives of the two sides in the case, Clarence Darrow and William Jennings Bryan, agreed on the grand import of the trial. It was not just about evolution, both insisted, but about control of public schooling. As Darrow pronounced angrily during the trial, "We have the purpose of preventing bigots and ignoramuses from controlling the education of the United States."[4] Bryan disagreed with Darrow's defense of Scopes but agreed that the real question at stake was who would control the public schools.[5]

Although many fundamentalists had rejoiced when Governor Austin Peay of Tennessee signed the Butler Bill into law in March 1925, none suspected the worldwide attention that Tennessee's law would soon draw.[6] The dramatic events of the Scopes "monkey" trial were set into motion when the American Civil Liberties Union (ACLU) offered to pay the expenses of a test case against the new law. Eager town boosters in Dayton, Tennessee smelled an opportunity to get some exposure for their hamlet and soon secured the cooperation of the young general-science teacher John Scopes. The trial still might have escaped undue attention had not William Jennings Bryan and Clarence Darrow offered to serve on the prosecution and defense, respectively. With the nation already eagerly following the "antievolution" crusade, the clash of titans provided the dramatic setting many Americans had been waiting for. With Bryan as their representative, many prominent fundamentalists hoped to promote the fundamentalist movement in the eyes of the nation and the world. James M. Gray of the Moody Bible Institute consulted with Bryan about his biblical arguments.[7] William Bell Riley, J. Frank Norris, and John Roach Straton all offered their services as well.[8]

Fundamentalists were right to be interested in the trial. It turned out to be the biggest single event that brought the fundamentalist movement to the attention of the American public. Many outside observers equated their stereotypes about rural Tennessee culture with the antievolution movement. Furthermore, many of those observers muddied the distinctions between the antievolution movement and fundamentalism as a whole. As we have seen, fundamentalism in the early 1920s was a diverse group of conservative evangelicals, and evolution opponents composed an even wider group. However, as a long-term result of such fervent scrutiny, fundamentalists felt intense pressure to conform their movement to the newly restricted popular stereotype. In spite of sustained efforts by some fundamentalists to build a fundamentalist movement that included a relatively wide coalition, the attention of the Scopes trial allowed outsiders, liberals, and even some fundamentalists to promote an image of rural, populist, anti-intellectual traditionalism as coequal with the entire fundamentalist movement.

This struggle had gone on throughout the decade. Recall liberal Baptist Harry Emerson Fosdick's 1922 attempt to marginalize fundamentalists as petty controversialists dedicated only to demonstrating "one of the worst exhibitions of bitter intolerance that the churches of this country have ever seen."[9] Fundamentalists of all stripes had countered these attacks for years by asserting that their movement did not start controversy, but merely acted in the mainstream spirit of Paul, "the true fundamentalist."[10] Fundamentalists, in the words of another defender, "are said to be causing division though in reality the cause of the division which threatens to divide the churches is the introduction of new views."[11]

The attention of the trial tipped the balance in the struggle to define fundamentalism. After the trial, liberals had a much easier time characterizing fundamentalists as ignorant aggressors. Fundamentalists after the trial had to scramble to deal with a rapidly constricting definition of their movement. They found that the trial had tested the scientific credibility of their movement with particular ruthlessness. Fundamentalists had entered the trial uniformly confident in the scientific superiority of their beliefs. However, the trial dramatically demonstrated the lack of scientific credibility of the movement's experts beyond the boundaries of fundamentalism itself. As with the other contentious aspects of fundamentalist identity, this crisis of scientific respectability eventually divided the fundamentalist movement. Some of the first fundamentalists slowly drifted away from the movement. They did not choose to identify themselves with a movement that lacked intellectual and cultural respectability outside its own borders. Others embraced their scientists more tightly, and worked to build an independent fundamentalist scientific establishment that no longer looked to mainstream science for legitimation. However, this shift in the struggle to construct viable boundaries for fundamentalism was not directly apparent. Fundamentalists and their enemies fought the struggle over fundamentalism's meanings and its scientific credibility over the course of the next several years.

Because of the simmering antievolution controversies around the country, the trial in Dayton, Tennessee, attracted intense media attention. Harold Odum, a well-known southern sociologist, estimated that the reporting from the "monkey" trial would have filled 900,000 print pages. He claimed that "no periodical or any sort, agricultural or trade as well . . . has ignored the subject."[12] Film crews shot miles of film and the *Chicago Tribune* engineered what it claimed to be the world's first radio broadcast of the event.[13] Reporters were not the only people flocking to town. John Scopes later remembered a circus-like atmosphere. In his opinion, "Ringling Brothers or Barnum and Bailey would have been pressed hard to produce more acts and sideshows and freaks than Dayton had."[14] Indeed, in addition to musicians and carnival-style attractions, evangelists and ideologues of all kinds rumbled down Main Street in their gaudy vehicles. One white supremacist from Georgia raised eyebrows from locals and out-of-towners alike when he rolled into town "in a bungalow on wheels, wearing an opera hat, an alpaca coat, and an ancient pair of trousers similar to those worn by policemen." To the Georgian's surprise, Tennesseans were not much interested in his long expositions on the similarities between monkeys and African Americans. They tended to be more concerned with turning the world's attention to profit.[15] To that end, local merchants eagerly embraced the monkey motif. One butcher shop put a sign in its window proclaiming, "We handle all kinds of meat except monkey." A drug store ran a similar advertisement in the local paper: "Don't monkey around when you come to Dayton but call us."[16]

Many of the reporters and correspondents from outside the region promulgated well-worn stereotypes of the local populace. One New York reporter described locals as primitives to whom "the devil still means a great deal." This reporter demonstrated the stereotype that liberals had been trying to graft onto the fundamentalist movement since the movement's birth in 1920. In the eyes of the reporter, Tennesseans, like fundamentalists everywhere, suffered in their isolation from "sophisticated communities." Like other nonelite groups, Daytonians were painted as cultural and intellectual throwbacks, motivated by "emotional state[s]" rather than what appeared to some liberals to be the self-evident logical and rational beliefs of urban progressives.[17]

Hostile urban commentators often expressed surprise at the realities of Dayton. H. L. Mencken, relentless foe of what he called America's small-town "booboisie," noted his surprise when he alighted from his train into a "country town full of charm and even beauty."[18] Another reporter from the segregated urban North expressed shock at the way "Negroes mingled freely with white persons" in Dayton. Apparently, that reporter had expected a Southern town to be ruled by conspicuous racial terror and feuding.[19] A New York correspondent had expected to find that the Butler Act was supported by only the most isolated and ignorant of rural Tennessee voters. Instead, he found widespread enthusiasm for the antievolution measure, including support from "bankers, farmers, merchants, lawyers and newspaper editors."[20]

Nevertheless, blinkered by their expectations, or perhaps hoping to sell more newspapers to northern urban audiences, most reporters from leading Northeastern newspapers perpetuated existing stereotypes of rural Tennesseans.

Reporters repeatedly characterized the crowds of locals who ventured into town during the trial in terms such as "hill people," "slow of motion and speech."[21] Press reports brimmed with condescending descriptions of "sober-faced, tight-lipped, expressionless" dirt farmers rolling into town to take part in the unusual spectacle.[22] H. L. Mencken sent home an account of his visit to a Pentecostal prayer meeting in the hills outside town, in which "pathetic" "half-wits" engaged in a "barbaric grotesquerie" of prayer.[23]

Much of the reporting included elements of sensationalism and cultural voyeurism, included as "local color." Henry Hyde, a correspondent from Baltimore, painted an exotic picture for his East-Coast readers of the religious geography of the small town. "Out in the woods on the mountains around Dayton the echoes of the wild shrieks and the hysteric moans of the Holy Rollers are just dying away," he wrote. Like Mencken, Hyde had recently visited a Pentecostal religious service outside of town. His description of local culture as somehow trapped in an ecstatic primeval past was common among out-of-town reporters.[24] Russell Owen, a New York correspondent, blithely described Dayton's "remote mountaineers" as having "simple" minds. In a condescending attempt "in a way to respect their point of view," the writer began by noting, "Prejudiced they certainly are, bigoted they may be." For writers like these, Dayton's inhabitants inhabited a lost world of religious intensity and intellectual stagnation.[25]

These correspondents at the Scopes trial drew upon a long tradition in American letters of dismissive portrayals of Southern Appalachian culture. Historian Henry Shapiro has argued that by 1890, many Americans from outside the region had become fascinated by what they imagined to be the "apparent reality of degradation and degeneracy in the mountain region." Readers interested in the region as a stronghold of "local color" eagerly read novels such as John Easton Cooke's *Owlet* (1878), which described southern "mountaineers" as "destitute of all ideas beyond the wants of the human animal in the state of nature." By the end of the nineteenth century, William Goodell Frost popularized a view of those mountain dwellers as "our contemporary ancestors." These degrading stereotypes had been challenged by the 1920s, by which time an early forerunner to the Highlander Folk School had opened to celebrate authentic Appalachian culture. Nevertheless, correspondents at the Scopes trial readily used such still-popular language to describe the antievolutionists at the Scopes trial.[26]

Some correspondents, like Henry Hyde and Russell Owen, used such stereotypes in reports that attempted to be fair and balanced, in spite of their condescension and ignorance of local culture. Other writers from outside the region made no pretense to sympathy or interest in the ideas and beliefs of locals. Maynard Shipley, journalist and president of the Science League of America, blamed the cultural retardation of East Tennessee for its support for the Butler Act. Many people, Shipley argued, assumed that "the enactment of the Tennessee antievolution law was to be accounted for by the backward condition of the commonwealth, where the percentage of illiteracy runs very high."[27] Liberal Tennessee native T. S. Stribling sought to bolster this stereotyped understanding of Tennessee's popular culture in his 1926 novel *Teeftallow*. In this fictional condemnation of an inadequate educational and cultural system, Stribling portrayed the backward condition of Tennessee mountain dwellers in the person of "Professor" Lem Overall. In one scene, Overall argued before a local court that evolution should be banned in all educational institutions. In his introduction to the court, Overall described himself as "a man of science an' as a representative of the edjercational intrusts of this county." As had other liberal writers, Stribling hoped to define East Tennessee culture as woefully behind the times.[28]

Just as East Tennessee was portrayed as a culturally primitive region, evolution supporters repeatedly equated fundamentalism and the antievolution movement with such stereotypes of regional southern Appalachian culture. One writer in the *New York Times*, for instance, described Scopes trial presiding judge John Raulston as "a product of the Tennessee hills, having practiced law and presided at court in the midst of this Fundamentalist atmosphere all his life."[29] Inveterate critic H. L. Mencken concluded from his experience in Tennessee that the Scopes trial "serves notice on the country that Neanderthal man is organizing in these forlorn backwaters of the land, led by a fanatic, rid of sense, and devoid of conscience."[30]

These stereotypes had a lasting political impact. Years after the trial, University of California president William W. Campbell used this equation of fundamentalism with stereotypes about Tennessee as a political weapon. In 1927, Campbell denounced California's proposed antievolution bill. Such a bill, he argued, would "in the eyes of the entire intellectual world place California with Tennessee and Mississippi as representative of the especially benighted parts of the United States."[31] According to one New York writer, California, although one of the states "least in sympathy with the spirit that prevails in Tennessee," had an "element that would hark back to antiquated ideas and to backward policies." In all these attributions, hostile stereotypes about the culture of East Tennesseans became a political and cultural shorthand with which to circumscribe the boundaries of fundamentalism.[32]

In this environment of intense spectacle and cultural confrontation, the trial itself began in a manner that many observers found dry, legalistic, and boring. Its first day was limited to the mundane business of reindicting Scopes—necessary due to doubts about the legality of the original grand jury indictment—and choosing a jury. Editors worked diligently to spice up the coverage. The

New York Times, for example, apparently hoped to spark interest by proclaiming in a frontpage headline: "Jury Includes Ten . . . One Is Unable to Read." Even when the trial itself could not hold readers' interest, editors apparently figured, perhaps indictments of Tennessee culture could.[33]

In any case, when the trial convened again on Monday, Clarence Darrow willingly provided some more dramatic fodder for reporters. As part of the defense strategy to prove the Butler Act unconstitutional, Darrow delivered a scathing indictment of the law. He repeatedly characterized the law's supporters as driven by "bigotry and ignorance." The law itself, Darrow argued, came from a "strange, weird, impossible and medieval" mindset.[34] If allowed to stand, Darrow concluded, the law would only encourage more "ignorance and fanaticism," both in Tennessee and around the nation.[35]

The crowded courtroom exploded with enthusiastic applause at the conclusion of Darrow's peroration. Clearly, in contrast to reporters' assumptions, the locals were as receptive as Americans anywhere to pleas for religious toleration and the greatness of the American way. Nevertheless, in this case Darrow's argument did not match contemporary mainstream constitutional thinking. In 1925, in Tennessee and the rest of the nation, the goal of the separation of church and state was very different from later interpretations. Government was traditionally prohibited from favoring any denomination or sect, but was not prohibited from encouraging Protestant religiosity in general. Not surprisingly, Judge John Raulston rejected the defense plea to dismiss the indictment.[36]

The defense had expected as much. Given the failure of their opening gambit, the defense team shifted to a second strategy. In the words of historian Jeffrey Moran, the defense now worked to "publicize the intellectual shortcomings of the Butler law and its supporters."[37] They planned to introduce testimony from scientific and theological experts. The two-pronged attack would show that evolutionary theory had become a mainstay of mainstream science and that evolutionary thinking did not necessarily conflict with Christian belief. In short, the defense team explicitly hoped to discredit their opponents' legal case by deflating the intellectual and cultural presumptions of the antievolution movement.[38]

Originally, the prosecution team had hoped to counter with expert witnesses of its own. By the time of the trial, however, the prosecution faced an embarrassing lack of experts. As a result, they were forced to argue against the use of expert testimony at all. Reluctantly, William Jennings Bryan made the case that this trial did not call for outside experts. Instead, Bryan contended, the narrow legal point was all that mattered. Tennessee's legislature had passed a law, and Scopes had broken it. Bryan had originally hoped to turn the trial into just such a showdown between experts. However, in order to win this case he appealed instead to the right of Tennesseans to make their own laws, heedless of outsiders' expert sentiment. At the close of the second week of the trial, Judge Raulston agreed.[39]

By ruling out the inclusion of evidence from scientists and theologians, the judge made the rest of the trial look to be a perfunctory affair. Many reporters—including H. L. Mencken—went home. In fact, however, the most dramatic and exciting scenes of the trial were still to come. On Monday, July 20, 1925, Judge Raulston moved the court outside. He was worried that the standing-room-only crowds had been putting too much pressure on the floor of the courthouse. Also, the unusually hot weather had produced an insufferably sweltering environment inside the brick courthouse. The judge hoped an outdoor grandstand would be safer and more comfortable for everyone. The new seating allowed more spectators to watch as the defense team called William Jennings Bryan himself to testify.[40]

The prosecution initially protested the defense move. There was little legal need for a prosecution lawyer to be questioned on the witness stand. Bryan, however, jumped at the chance. This direct confrontation would give Bryan the opportunity he had been looking for. If he could question the defense witnesses in turn, Bryan felt confident he could demonstrate the dangers of evolutionary theory and the intellectual weaknesses of its defenders.[41]

The defense hoped to use Bryan's testimony to discredit the intellectual foundation of fundamentalism and the antievolution movement. If even Bryan could be shown to interpret the Bible, the argument that evolution necessarily shattered traditional Protestant faith could be toppled as well. In other words, if all Christian belief required some measure of interpretation of biblical text, then neo-Darwinism's challenge to the Genesis creation stories would seem less threatening. In the words of defense attorney Arthur Garfield Hays, if Bryan admitted that even he interpreted the Bible, "He must have agreed that others have the same right." Even better, if Bryan could be shown to be ignorant of key scientific arguments, the entire fundamentalist movement could be dismissed more easily as ignorant and intellectually isolated.[42]

Bryan knew the danger. In his testimony, he cautiously pointed out that he did not believe in the literal truth of the entire Bible. At times, the Bible spoke figuratively, Bryan argued, such as in saying "Ye are the salt of the earth." But this did not, according to Bryan, open the door for such wide-ranging interpretations as those favored by liberals. For instance, Bryan steadfastly defended the supernatural premise of miracles. When questioned about the story of Jonah and the whale, Bryan articulated one of the core principles of fundamentalist epistemology. In the story, a sinful Jonah had been swallowed whole by a "big fish," and spent three days inside. To the minds of many liberals, this physical impossibility weakened the ability of many to believe in the truth of the Bible. It demonstrated to liberals the mythic nature of this collection of ancient writings. To Bryan, however, knowledge began with the Bible. If the Bible reported something as true, then that truth should be the starting point for investigation. Fundamentalists like Bryan did not find any difficulty in reconciling the physical impossibility of many miracles with their historical veracity. As Bryan taunted Darrow from the witness stand, the story of Jonah and the whale "is hard to believe for you, but easy for me. A miracle is a thing performed beyond what man can perform. When you get beyond what man can do, you get within the realm of miracles; and it is just as easy to believe the miracle of Jonah as any other miracle in the Bible." His testimony on points like this one demonstrated the cultural divide between the two sides. For Darrow and his ilk, Bryan's willingness to believe the Bible regardless of its scientific dubiousness proved Bryan's ignorance. To the fundamentalist audience, Bryan's retort proved the shameful and willing ignorance of skeptics such as Darrow.[43]

Bryan also defended the widest possible cultural identity for fundamentalism. In the face of Darrow's accusation that Bryan pandered only to "bigots and ignoramuses," Bryan sought to show that fundamentalists included a much more intellectually respectable group. He blasted Darrow for insulting the locals as mere "yokels."[44] Further, he objected to Darrow's use of the word "prejudices" to describe Bryan's belief. "I don't think," Bryan retorted, "I am any more prejudiced for the Bible than you are against it." Bryan was unwilling to yield to Darrow's attempts to circumscribe the definition of fundamentalism within narrow boundaries. Yes, Bryan tacitly acknowledged, a significant gulf existed between Bryan's and Darrow's beliefs. But Bryan would not accept that his own beliefs were those of a bigoted, isolated few. Rather, he hoped to use his testimony to show that the beliefs of fundamentalism could be defended calmly, intelligently, and successfully.[45]

The trial itself concluded the day after the confrontation between Darrow and Bryan. At the insistence of the prosecution, Judge Raulston expunged Bryan's testimony from the record. Conceding certain defeat, Darrow asked for an immediate jury ruling. Darrow hoped to move quickly to an appeal. The jury took only nine minutes to return with a guilty verdict. Judge Raulston assigned a minimum fine with the consent of all concerned. Unfortunately for the defense, only the jury had the legal ability to assess a fine, so the verdict was eventually dismissed. The trial never became the wider constitutional test the ACLU had hoped for.[46]

In the immediate aftermath of the trial, both sides claimed victory in various ways. In sharp contrast to later representations of the trial such as the play and movie *Inherit the Wind*, fundamentalists lauded Bryan's masterful performance on the witness stand. He had, in their eyes, successfully articulated and defended their beliefs. Norris, for instance, ran a banner headline in his magazine, reporting that "Bryan Wins Greatest Victory of his Career—Bible Triumphs Over Infidelity: Commoner Outwits Darrow in Dayton Evolution Trial."[47] Liberals saw Bryan's testimony in a very different light. One New York correspondent reported that the most remarkable thing about Bryan's testimony was that "there was no pity for his admissions of ignorance of things boys and girls learn in high school, his floundering confessions that he knew practically nothing of geology, biology, philology, little of comparative religion, and little even of ancient history."[48] As the more perspicacious reporter Russell Owen noted, "Each side withdrew at the end of the struggle satisfied that it had unmasked the absurd pretensions of the other."[49]

That both sides witnessing similar events could interpret them in such diametrically opposite ways demonstrates the profound division at work in the trial. Given their presumptions, however, the conclusions of liberal observers made sense. Most agreed that the "Remote Mountaineers" responsible for the antievolution campaign could have no influence outside of isolated hamlets like Dayton, Tennessee.[50] One secular writer confidently asserted the fundamentalists' inevitable disarray after the trial. "The mischief they can do is temporary and essentially ludicrous," the writer opined. "Mr. Bryan and his followers may bray forth the claims of ignorance and buttress their prejudices with a conveniently mediaeval theology, but their sophistries cannot endure long nor do much damage." The writer considered his prediction too self-evident to require proof. From the liberal perspective, any views so profoundly distinct from liberal opinion must quickly wither and die in the modern age.[51] Journalist and popular historian Frederick Lewis Allen soon enshrined this liberal conclusion as historical fact in his 1931 history, *Only Yesterday*. In this work, Allen reported that in spite of the technical legal victory for the fundamentalists, "Fundamentalism had lost."[52]

These predictions of fundamentalist disarray became so ubiquitous in the years following the trial that some enemies of fundamentalism found themselves unable to rally support. One secularist writer correctly warned that the forces of fundamentalism had become more aggressive after the "thrill of victory that they got from the Scopes verdict."[53] In 1927, Maynard Shipley, president of the Science League of America, sought desperately to warn his fellow secularists of this continuing fundamentalist activism. He bemoaned the "shame of Tennessee"[54] and warned that "the armies of ignorance are being organized, literally by the millions."[55] In the next year, the proevolution activist felt like a Cassandra with his repeated warnings. "Although it is continually asserted," Shipley lamented, "that 'the Dayton trial and Bryan's death ended the Fundamentalist drive' . . . this is far from being the case."[56]

Other prominent secularists agreed. In the aftermath of the trial, Mencken reflected dourly on his experiences as a correspondent for the *Baltimore Sun*. Mencken continued to attack Bryan even after his untimely death in the days following the trial, as a "charlatan, a mountebank, a zany without shame or dignity."[57] Bryan, in Mencken's opinion, had proven himself a dangerous "dog with rabies" at the trial. Bryan's followers were nothing more than a contemptuous gaggle of "yokels," a "forlorn mob of imbeciles."[58] Nevertheless, Mencken concluded, the fundamentalists had not been decisively defeated by the trial. "Heave an egg out of a Pullman window," Mencken concluded morosely, "and you will hit a Fundamentalist almost anywhere in the United States to-day. They swarm in the country towns, inflamed by their shamans, and with a saint, now, to venerate. They are thick in the mean streets behind the gas-works. They are everywhere where learning is too heavy a burden for mortal minds to carry, even the vague, pathetic learning on tap in little red schoolhouses." Worst of all, to the cynical Mencken, there was no doubt that the fundamentalists would win. Americans tended to lean toward the stupid rather than the enlightened, Mencken believed. They would soon embrace such imbecility, rather than building a more rational American society.[59]

For their part, fundamentalists moved quickly to capitalize on what they saw as an epochal victory. Bryan's death a few days after the trial only heightened this sense of momentousness. Most fundamentalists believed the trial had afforded them a unique opportunity to press their case. However, in the aftermath of the trial a few fundamentalists publicly challenged the wisdom of the antievolution campaign. If it had generated so much negative attention at the Scopes trial, some astutely believed, it could only lead to pressure on fundamentalists to accept or reject a newly restricted popular image of fundamentalism. This nervousness on the part of some fundamentalists proved prescient. However, immediately after the trial, it was also relatively rare. Few leading fundamentalists had the temerity to question the achievements of the trial. One of those holdouts was Curtis Lee Laws, a Baptist fundamentalist who had coined the term fundamentalism in the summer of 1920. Writing in the immediate aftermath of the trial, Laws observed glumly, "The Scopes trial ought never to have been made an issue of fundamentalism. In our opinion the fundamentalists will be wise to major on other matters than evolution." Laws worried that the fundamentalist movement would be subsumed by popular stereotypes about the antievolution movement in the wake of the trial. Laws also hoped that the fundamentalist movement would continue to welcome a relatively wide range of opinion about the teaching of evolution. Before the trial, after all, many leading fundamentalists had welcomed the teaching of evolutionary theory, if done properly.[60]

In the immediate aftermath of the trial, however, most fundamentalists did not share Curtis Lee Laws' judicious worry. Instead, they voiced their excitement at the great cultural victory of the trial, and rushed to take advantage of what they viewed as the outpouring of positive publicity from the trial. Riley immediately announced plans to move forward with one of his pet projects: a "Fundamentalist University." Indeed, Riley asked his readers, "What memorial could match in fitness a great Fundamentalist University, erected in [Bryan's] memory and destined to wear his name while time should last?"[61] The leaders of the Bible Institute of Los Angeles (Biola) wholeheartedly supported the idea for a Bryan University, either in Illinois, as Riley had suggested, or, as Florida fundamentalist George Washburn suggested, in Dayton itself. In any case, the Los Angeles fundamentalists made clear their enthusiasm for Bryan's recent victory in the trial itself. When witnesses saw Bryan's testimony, Biola editor T. C. Horton reported, they saw "a real man, face set like a flint, eyes lighted with a radiant glow, standing four square upon his feet, with Bible in hand, ready to fight for it, ready to die for it, for he had tested it, proved it, believed with all his heart that it was worth defending to the death."[62] Phillip E. Howard, who eventually became editor of the fundamentalist *Sunday School Times*, reflected on Bryan's probable feelings during Darrow's examination. Howard compared Bryan to the British at the Second Battle of Ypres in 1915 when they "caught their first intimation of poison gas." In Howard's opinion, Darrow's treacherous aggression could not shake Bryan from his heroic and ultimately successful defense.[63]

This widespread sense of victory and momentousness led many fundamentalists to increase their efforts on behalf of antievolution legislation. Due in part to their efforts, the most turbulent year in state legislatures around the nation was 1927, not 1925. In that year, fourteen state legislatures, plus the United States Congress, seriously considered some form of fundamentalist legislation. As Chapter 6 will demonstrate, those laws often decreed bolder, more sweeping changes in educational policy than earlier laws had dared.[64]

In spite of the ambivalence of thoughtful fundamentalist observers like Curtis Lee Laws, most fundamentalists took heart from the Scopes trial and from Bryan's defense of fundamentalism's basic intellectual presuppositions. It would only be in the years to come that more fundamentalists would find themselves wrestling with the trial's stereotyped image of fundamentalism. As the publicity from the trial made clear, many outside observers equated the entire fundamentalist movement with extant popular stereotypes about Tennessee "mountaineers." In the later years of the 1920s, liberals would use this stereotype as a convenient weapon with which to discredit the entire fundamentalist movement. As we shall see, this tactic forced fundamentalists to either embrace or reject the newly restricted boundaries of fundamentalism.

These developments, however, played out only gradually and haltingly in the coming years, as liberals and fundamentalists engaged in the complicated struggle over the meanings of fundamentalism. One of the areas in which the Scopes trial contributed most profoundly in this fight was in the attitude of fundamentalists toward mainstream science. In general, fundamentalists had long considered their movement to be in the vanguard of mainstream American science. They had not recognized the important changes that had occurred in the scientific establishment since the end of the Civil War. Most fundamentalists entered the Scopes trial confident of their ability to trounce their scientific foes in open debate. Events of the trial demonstrated the gulf between fundamentalists' ideas about science and those of mainstream American scientists. For some fundamentalists, this discovery began a slow process of disengagement from the fundamentalist movement. For others, it demonstrated the superiority of fundamentalist science and heralded the need for an independent fundamentalist scientific community, immune from the criticisms of those who did not share the epistemological presuppositions of the fundamentalists.

From the beginning of the school controversies of the 1920s, both sides had claimed the mantle of science and expert knowledge. This was not unique to this educational battle; similar fights had raged for decades over issues such as the teaching of "Scientific Temperance."[65] In the case of the fundamentalist school campaigns, however, both contemporary critics and later historians often assumed that the fundamentalists opposed science, since they so vehemently disputed the teaching of evolution. It has also been widely presumed that the fundamentalists were hostile to the contemporary cult of expert opinion, since the stereotypical educational expert was thought to favor the teaching of evolution and the secularization of public schools.[66]

In reality, fundamentalists of the 1920s valued the prestige of science and eagerly sought to buttress their arguments with the testimony of acknowledged experts. However, fundamentalists' conceptions of science were generally distinct from ideas that had become mainstream by the 1920s. Historian George Marsden, in his authoritative study of the intellectual roots of twentieth-century fundamentalism, has identified the reverence with which fundamentalists regarded science. As Marsden demonstrated, fundamentalists in the 1920s commonly viewed science from a traditional nineteenth-century perspective. In this understanding, true science consisted of the classification of facts according to an overarching scheme. Unlike the new approach to science typified by Darwinism, fundamentalists believed true science, using traditional assumptions, did not rely on hypotheses and theories, but rather carefully generalized and classified facts beginning with an authoritative truth.[67]

While fundamentalists argued for the superiority of their understanding of science, however, they did not argue that they wanted to return science to its nineteenth-century status. Instead, fundamentalist activists usually argued that their scientific understanding included the latest discoveries of mainstream science as well. It was only the prejudices of mainstream scientists, many fundamentalists argued, that prevented those mainstream scientists from understanding the significance of these cutting-edge discoveries. Before the Scopes trial, most leading fundamentalists assumed they could convince mainstream science of the superiority of the fundamentalist view. After the trial, fundamentalist scientists slowly acknowledged that their views, though superior to those of mainstream science, would never be allowed to compete with the views of those who had accepted the claims of organic evolution.

Due to this difference in their understandings of science, both fundamentalists and their foes claimed to be championing the cause of true science during the educational controversies of the 1920s. Thus, when fundamentalist Riley met atheist Charles Smith in a public debate over whether evolution should be taught in public schools, Smith asserted that true science involved hypothesis, experiment, and gradual verification. Riley countered that true science was only "knowledge gained and verified," not a group of developing hypotheses.[68] Fundamentalist evangelist Martin agreed. He explained the traditional view that *Evolution is not science. Face the facts: Science is knowledge, classified knowledge.*"[69]

For many fundamentalists, the basic error of their opponents lay in the misunderstanding of this central scientific truth. In the traditional understanding, one began with the authority, and sorted facts according to this given pattern. For fundamentalists, this source was God, and His Truth was revealed to humanity in the Bible. Unlike the assumption of many of their foes, fundamentalists did not simply ignore the truth if it contradicted the Bible. Instead, many fundamentalists began with the scientific presupposition that the Bible was the authority, and apparent contradictions to the Bible must only be errors of observation or interpretation. Fundamentalist authors often drew upon the traditional "two books" argument. According to this traditional argument with roots back to the earliest Puritan-era scholastic debates, God had created two books for humanity to study. The first was nature; the second was the Bible.[70]

In the words of fundamentalist author and pollster Alfred Fairhurst, "There can be no conflict between true science and true religion, for God is the author of both."[71] George McPherson, evangelist and author, similarly argued, "Ours is not a conflict between science and the Bible, for true science is the hand-maid of Christianity."[72] A third writer tried to clarify the fundamentalist harmony between science and religion: "True science, which is true knowledge of the universe or the facts of nature, cannot contradict the Bible, because God would then be contradicting Himself."[73]

William Jennings Bryan approached these issues with particular zeal. He acquired a reputation among nonfundamentalists as an opponent of science with his claim that "it is better to trust in the Rock of Ages, than to know the age of the rocks." Opponents interpreted this statement as an affirmation of the closed-mindedness of fundamentalists and their inability to see truths that contradicted the Bible. Instead of a dismissal of science, however, Bryan's famous line must be understood within the framework of his definition of science. One should not begin with geologic data, Bryan believed, but with the Bible, the authoritative guidebook that explained scientific discoveries. "Science," according to Bryan, "is classified knowledge. . . . Darwinism is not science at all; it is guesses strung together. There is more science in the twenty-fourth verse of the first chapter of Genesis . . . than in all that Darwin wrote."[74] Hardly an opponent of science, Bryan cultivated a public image as a devoted amateur scientist. He insisted that he was fighting on the side of true science. He even joined the American Association for the Advancement of Science (AAAS), a notoriously proevolution group. Bryan refused to allow this group to monopolize science and he maintained his membership despite open hostility from some other members.[75]

Before the Scopes trial, Bryan believed vehemently that the scientific establishment was on his side, or at least that it soon would be. His beliefs, Bryan argued in 1922, did not oppose science, but rather included science's latest discoveries. Soon, mainstream science would reject organic evolution, since, as Bryan noted with some justification, "natural selection is being increasingly discredited by scientists."[76] Of course, to most contemporary scientists, this was an unfair claim. Bryan had noted the arguments among scientists about the method of evolutionary change. In 1922, American scientists overwhelmingly accepted the validity of evolution, even as most questioned Darwin's suggested method of transmutation. Bryan took advantage of such disputes within mainstream science to buttress his belief that his version of biblically based antievolutionism could find support from prominent scientists.[77]

Up to the time of the Scopes trial, Bryan mistakenly believed that his views on evolution could be backed up by a veritable army of reputable scientists. As the prosecution team readied its case for the Scopes trial, Bryan confidently promised his fellow attorneys that he could bring "many of the leading scientists" to help their case.[78] These scientists, Bryan promised, would show the world "that our side was prepared to hold its own against their committee of scientists."[79] He also steadfastly believed, before the trial, that his expert witnesses would lay out a convincing scientific case against evolution. As he wrote to one sympathetic correspondent, "I am expecting a tremendous reaction as a result of the information which will go out from Dayton."[80]

Other fundamentalists shared this belief. As Norris wrote Bryan excitedly in the days leading to the trial, the chimerical panel of experts for the prosecution were "real scientists and could meet any hoax or fraud that might be made by the defence."[81] This had long been assumed by many of the leaders of the fundamentalist movement. In general, early fundamentalists did not ignore the fact that their scientific beliefs contradicted those of many mainstream scientists. Fundamentalists generally assumed, however, that many mainstream scientists had entered a methodological blind alley. Prominent fundamentalist editors and pundits repeatedly reassured their readers and themselves that their scientific beliefs were based not only on their religious presuppositions but also on the most "Modern Scientific Discoveries."[82] In the years leading up to the Scopes trial, fundamentalist readers often read that evolution would soon be overturned as part of the scientific establishment by "recent discoveries in geology."[83]

Fundamentalists before the Scopes trial presumed that the rigors of science would soon bring erring mainstream scientists around to agreement with fundamentalist scientific presuppositions. Leading fundamentalist scientist George McCready Price argued tirelessly during the years leading to the Scopes trial that mainstream science would soon find itself discredited in regard to evolution. As James M. Gray of the Moody Bible Institute reported, Price had proven that "the progress of science is destroying much that till lately passed for gospel" among evolutionists.[84] Price himself privately reassured William Jennings Bryan before the Scopes trial that leading evolutionists were "out of date,—behind the times,—and don't know it."[85] Leading Lutheran fundamentalist intellectual Leander S. Keyser dared nonfundamentalist scientists to accept his challenge and return to a truer fundamentalist approach to science. "Why not be scientific by accepting the evident and wholly adequate account of ultimate causes and sources?" Keyser asked. "Why indulge in remote and misty speculations, and then call them by the noble name of science?"[86]

Much of this belief rested on selected expert testimony from writers claiming mainstream scientific credentials. Many historians have agreed with Jeffrey Moran that that the fundamentalist movement formed part of "an older democratic ethos" that opposed "the rising authority of experts in American culture."[87] In the first years of the movement, however, this was not the case. Fundamentalists clung as dearly as any other Americans to the belief that expert testimony provided the key to solving social and political problems. Most went to extreme lengths to demonstrate the widespread support of experts for their positions. Martin, for instance, noted with pride that he could name 120 experts to defend his arguments in *Hell and the High School.* In that book, sixty-seven pages out of a total of 175 simply listed experts, and described their credentials.[88] Other fundamentalist authors cited similarly cumbersome amounts of expert testimony to bolster their claims. Alfred Fairhurst, on one typical page in *Atheism in Our Universities*, included only twenty-three original words. The other 107 introduced "leading writers on evolution" and quoted them at great length. Fairhurst, like Martin and other fundamentalist leaders, did not appeal to any populist disdain for experts. Rather, fundamentalists scrambled to cite as many high-powered experts as they could.[89]

Unfortunately for many early fundamentalists, the pool of experts from which fundamentalist writers could draw tended to include only other prominent fundamentalists. Fundamentalist writers often inflated one another's scientific credentials. Riley, for instance, in his debate with the atheist Charles Smith, included a list of experts widely shared among antievolution writers. Riley listed Alfred Fairhurst as one of his premier scientific experts.[90] Fairhurst had indeed written widely on evolution and education. However, he was hardly an eminent scientific expert. Fairhurst had taught natural science for several years at Transylvania University in Lexington, Kentucky, but he had completed only a single year of graduate work. Even this obscure career ended in a humiliating dismissal from classroom teaching.[91] Riley's other scientific experts, such as nineteenth-century geologist Joseph LeConte, had indeed been eminent scientists who had questioned Darwin's theory of natural selection. However, even LeConte accepted theistic evolution. Furthermore, LeConte's work had been completed before most mainstream scientific opinion had turned decisively in favor of the theory of evolution.[92]

Fundamentalist scientist Arthur I. Brown similarly assured fundamentalist readers that evolutionary ideas "have long since been discarded by scientific leaders." Brown lamented in 1922 that mainstream scientists mistakenly ignored the objections of,

world-renowned men like Virchow of Berlin, Dawson of Montreal, Etheridge of the British Museum, Groette of Strassburg University, Paulson of Berlin, Clerk Maxwell, Dana, Naegeli, Holliker, Wagner, Snell, Tovel, Bunge the physiological chemist, Brown, Hofman, and Askernazy, botanists, Oswald Heer, the geologist, Carl Ernst von Baer, the eminent zoologist and anthropologist, Du Bois

Reymond, Stuckenburg and Zockler, and a host of others. . . . It seems to be a fact that NO opinion from whatever source, no matter how weighty or learned, is of any account with those who are consumed with the determination to reject the Bible at any cost, and shut God out of His universe.

As had other lists of fundamentalist scientists, Brown's suffered from anachronism and sloppy analysis.[93] Like Joseph LeConte, many of Brown's experts had indeed been prominent scientists who had questioned evolutionary theory. However, most, like John William Dawson, had completed their work in the nineteenth century. Even Dawson during his lifetime was acknowledged as a unique hold-out against the premises, if not the details, of organic evolution. And even Dawson, like LeConte, accepted the possibility of theistic evolution.[94] Other nineteenth-century scientists on Brown's list, such as James Dwight Dana, had eventually accepted the consensus view in favor of evolution.[95] Nevertheless, for Brown, the dismissal of this impressive-sounding list of scientific experts could only result from a willful ignorance of scientific verity.

In the days leading to the Scopes trial, most fundamentalists assumed that their lists of experts could provide a battery of unimpeachably "expert" scientific testimony at the Scopes trial. Thoughtful fundamentalists could not help but be surprised when none showed up. This failure caused Bryan and many other fundamentalists to question their earlier assumptions about the relationship between their own ideas about science and those of mainstream American scientists. After all, Bryan had earnestly requested help from his scientific and theological connections. By the time of the trial, however, only one supporter with any mainstream scientific credentials had volunteered to come. Dr. Howard A. Kelly, a gynecologist at Johns Hopkins University, accepted Bryan's invitation to testify for the prosecution. However, Kelly warned Bryan that Kelly had accepted the evidence for evolution of lesser animals. Worried that such testimony might weaken the legal case, Bryan put Kelly on standby status.[96]

Bryan was more bitterly disappointed by George McCready Price's lukewarm attitude toward the trial. By the time of the trial, Price had become the leading scientific expert of the fundamentalist movement. Many fundamentalists, Bryan included, accepted without question Price's claims that his work was in the cutting edge of mainstream American science. Bryan had hoped to use Price's testimony as the centerpiece of his scientific testimony. In spite of Bryan's eager plea for help, Price was teaching in England at the time of the trial. Price was not willing to return to the United States for the trial, nor was he overly enthusiastic about the prospect of the trial.[97]

Even worse than Price's absence, the trial brought fundamentalist attention to the fact that Price utterly lacked credibility among nonfundamentalist scientists. One of the shocks of the Scopes trial came when Bryan cited Price as his scientific expert during his examination by Darrow. Darrow smelled his chance and pounced on Price's scientific reputation. "[Bryan] has quoted a man that every scientist in this country knows is a mountebank and a pretender and not a geologist at all," Darrow scoffed.[98] In fact, Price's scientific credentials were very limited. His formal scientific training included only a few "elementary courses in some of the natural sciences."[99] Price had failed to publish any of his findings in peer-reviewed scientific journals.[100] His foundational belief in the Genesis account of creation, supported by the writings of Seventh-day Adventist prophet Ellen G. White, no longer had any support among mainstream scientists. The events of the Scopes trial made clear to many fundamentalists that mainstream science would not soon accept Price's claims. As many fundamentalists began to recognize, American scientists and research universities had whole-heartedly, if mistakenly, embraced the presuppositions of organic evolution.[101]

In addition to grafting an inescapable stereotype of southern Appalachian hillbilly culture onto the meanings of the fundamentalist movement, the crisis of the Scopes trial challenged fundamentalists to acknowledge that their scientific experts had no credibility beyond their own movement. For some fundamentalists, this shock contributed to their desire to distance themselves from the fundamentalist movement as a whole. If the movement could not win intellectual and cultural respect beyond its own boundaries, many of the first fundamentalists were willing to separate themselves from it. Other early fundamentalists, however, reacted to this crisis very differently. Many fundamentalists concluded that this lack of mainstream scientific respectability called the reliability of mainstream science into question. These fundamentalists continued to support self-declared fundamentalist scientific experts. Eventually, as we will see in Chapter 9, this small group of fundamentalist scientists led a successful long-term campaign to create a sustainable scientific community of their own.

PART III

Monkeys and Modernism

CHAPTER 6

School Legislation after Scopes

In the aftermath of the Scopes trial, many fundamentalists reeled from the very public attacks on their intellectual credibility. Even before the trial, fundamentalists had been surprised at the vigorous opposition to their school laws. In the years following the Scopes trial, fundamentalists tried a variety of strategies to cope with this surprisingly strong opposition. Some advocated even stronger and more sweeping school laws. Others pressed for more narrowly focused antievolution laws. In both cases, fundamentalist-backed school laws met with mixed success in the later years of the 1920s. Failures often dispirited fundamentalists and convinced them that they did not enjoy the mainstream support many had expected. Even successes often lent increased support to the new stereotype about fundamentalism and the antievolution movement.

After the Scopes trial, much of the same campaign continued under the banner of the fight for "antievolution" laws, even though many of the proposed laws staked a much broader claim on the content of public schooling. As they had before the trial, fundamentalist activists pressed their belief that evolution was only one of the many kinds of pernicious teaching that could seduce impressionable young minds away from evangelical faith. Yet they also continued to use evolution as a symbol of this wider web of educational dangers.

For instance, roughly a year after the conclusion of the Scopes trial, the Supreme Court of Tennessee heard an appeal of Scopes' guilty verdict. Far from a narrow argument about whether or not John Scopes had taught from a textbook containing evolutionary theory, the prosecution in this appeal insisted that the issue ranged far beyond the teaching of evolution. Lawyers for the state emphasized that "the Tennessee legislature passed [the Butler Act] to stamp out worse things" than evolution. It was more about protecting students against creeping communism and social anarchy than with Darwin or natural selection.[1] Indeed, the prosecution averred that one of the primary goals of the "antievolution" Butler Act had been to "preserve the Bible" in the public schools.[2]

Many fundamentalists shared this broad vision of their antievolution activism; and although the Scopes trial soured some leading fundamentalists on the prospect of the antievolution struggle, many others continued and even intensified their campaigns to support such legislation across the country.

In all these fights, fundamentalists and their foes operated with a keen awareness of the impact of the Scopes trial on the popular image of fundamentalism. It would be overly simplistic to assert that these struggles over the nature of fundamentalism determined the success or failure of the legislative campaigns. Every city and state included a different mix of personality, population, and culture. Local activism or history usually determined whether or not fundamentalist school rules could win a critical mass of political support. Nevertheless, as local school boards and state legislatures debated the propriety of teaching evolution and other doctrines objectionable to fundamentalists in the years following the Scopes trial, the new stereotype of the fundamentalist movement often played a key political role. Liberals often used the new stereotype to ridicule antievolution laws as the last resort of ignorant reactionaries. Fundamentalists often rallied political support for a positive image of the new stereotype: uncompromising Southern defense of traditional values.

In early 1926, Congressional representatives in the nation's capital debated the issue. Representative Thomas L. Blanton of Texas deftly maneuvered through parliamentary roadblocks in order to introduce the topic of conservative Protestant religiosity in public schools to the floor of the U.S. House of Representatives. Blanton's defense of the Congress's 1924 proviso banning Washington, DC, teachers from any curricula that taught "disrespect for the Holy Bible" echoed many of the themes from the Scopes trial. As Bryan had, Blanton argued that teachers should not be "turn[ed] loose upon unsuspecting pupils" to "teach any kind of doctrine they want." In this post-Scopes political and cultural environment, however, Blanton felt obliged to add, "I know the newspapers make fun of us and call us 'fundamentalists' whenever we want to inquire into what the children are being taught."[3] In spite of his defensive attitude about fundamentalism's new public meanings, Blanton and other supporters of the 1924 federal law eagerly jumped on the bandwagon of the new fundamentalist image. Blanton hoped to keep political pressure on the cultural politics of education, even though his pro forma amendment had no chance of changing any law.[4]

Blanton's short speech demonstrated the transformation wrought by the Scopes trial and its attendant publicity. Whereas Representative John William

Summers's 1924 amendment had sailed through the House without a whisper of protest, Blanton's 1926 statement generated heated and relatively lengthy debate. Representative Frederick Lehlbach of New Jersey received applause when he argued that such rules ought to be subject to the full legislative process. Lehlbach protested that such a rule forced a "most onerous and difficult duty" upon teachers and should not be enacted lightly.[5] Fiorello LaGuardia, representing his East Harlem district at the time, launched into one of his trademark tirades against the already dead amendment. LaGuardia insisted that students were "safe in the schools; they are learning to think." He blamed the "hysteria" and the "wave of intolerance" of recent fundamentalist school campaigns for enacting "ridiculous" school laws. LaGuardia hoped that the U.S. Congress would not "follow the mistakes or foolish conduct of any State legislature." In short, LaGuardia insisted that fundamentalists and their allies had misunderstood the nature of education. As did many liberals, LaGuardia believed that education must first serve to train skeptical, inquiring minds. He believed that fundamentalist school laws, at least as caricatured in the cultural environment of the Scopes trial, could only block education, not reform it.[6]

Intensely aware of such attacks, state legislatures continued to debate fundamentalist school laws.[7] In early 1926, the Kentucky state legislature heard and quickly buried in committee an antievolution bill modeled on that of Tennessee.[8] Fundamentalist activists had more luck in Louisiana. In May of that year, a series of so-called antievolution bills worked their way through the state House and Senate. The first House bill simply required the prohibition of "the teaching of evolution in all the universities, normal and other public schools and State institutions in the State of Louisiana." The short bill never came to a vote, nor did the more ambitious House Bill number 208. The aggressive bill, introduced by self-identified fundamentalist Representative Sambola Jones, hoped to ban the employment of any atheist in any capacity in Louisiana's public schools, as well as to bar the use of any school property by atheists. Later that month, a more limited bill to prohibit any teacher in any school or college from "teaching that mankind either descended or ascended from a lower order of animals" passed the House of Representatives, only to be stymied in the upper house. Yet another bill met a similar fate. Louisiana House Bill number 314 passed by a margin of sixty-four to seven in the state House. It did not explicitly target the teaching of evolution. Rather it banned the teaching of "anything which is subversive of the creed, faith, doctrine or belief held by any pupil or which gives a preference to or discriminates against the church, sect or denomination to which he belongs." In spite of its broad support among state representatives, state senators stalled it fatally in June.[9] Instead of accepting defeat, the State Baptist Convention continued to pressure the state board of education. Due to

Baptist activism, the state superintendent instructed parish school superintendents "to instruct their science teachers . . . to omit . . . pages" from Hunter's
Civic Biology (1914), the same book that had been at issue in the Scopes trial.[10]
After the Scopes trial, however, Louisiana science teachers would not have
to omit much. After such intense fundamentalist scrutiny, the book's publisher deleted a critical six-page section. In addition, the word "evolution" itself
was completely eliminated and the concept was treated much more equivo-
cally.[11] Other publishers struggled to come up with books that would not
attract unwelcome attention from antievolution activists. Henry Holt took out
three chapters that treated the evolutionary descent of humans from Truman
J. Moon's *Biology for Beginners* (1922). Like other major publishers including
Macmillan, Holt wanted the new one to be a "Texas" edition, attractive to any
district in which evolution had become a contentious issue. Some publishers
merely deemphasized the content that included theories of human evolution.
They removed pictures of Darwin from books, and took the word evolution out of their indices, but kept their content roughly the same.[12] At least
one prominent textbook publisher contacted prominent fundamentalists to
secure their stamp of approval for science textbooks. A representative of Ginn
and Company asked William Jennings Bryan to clarify his position that the
teaching of evolution as a theory, not as a fact, met with his approval. Bryan
refused to give his official seal of approval, and he encouraged the publisher
to produce some higher-quality antievolution school materials, in recognition
of the popular appeal of fundamentalism.[13]

After the Scopes trial, then, many American schoolchildren would not
encounter the language of evolutionary theory in their science textbooks, even
without antievolution school laws on the state books. Nevertheless, state legislatures continued to debate new antievolution laws. In 1926 Mississippi followed
Tennessee in enacting a law very similar to Tennessee's Butler Act.[14] The most
turbulent year in state legislatures, however, was 1927. In that year, fourteen
state legislatures debated antievolution laws. Legislators defeated some of the
laws quickly, but others, such as in Florida and Arkansas, passed in one of the
houses of the legislature before being narrowly defeated in the other. For both
supporters and opponents of these laws, however, simply having them debated
in North Carolina, South Carolina, Florida, Alabama, Arkansas, Oklahoma,
Missouri, West Virginia, Maine, New Hampshire, Minnesota, North Dakota,
Delaware, and California represented a surge of popular support for the fundamentalist cause. Even as the pace of such state legislation slowed down in coming years, with only one state, South Carolina, reconsidering a law in 1928 and
only Texas in 1929, liberals and fundamentalists alike reported a wave of public enthusiasm for such laws.[15] Fundamentalists trumpeted their new activism

"all across the continent" in favor of "antievolution" school laws.[16] Their opponents indulged in similar exaggeration. The American Association of University Professors (AAUP), one of the staunchest enemies of antievolution laws, announced at the beginning of 1927 that seventeen states would consider such legislation in the coming year, although fewer states actually did so. Like their fundamentalist opponents, the AAUP assumed a rising tidal wave of political support for such laws.[17]

During this crest of fundamentalist legislative activism, many state legislators responded with alarm to antievolution activism in their own states. They accepted unquestioningly the stereotyped image of fundamentalism in the wake of the Scopes trial. In Maine, for example, lawmakers debated whether to relegate an antievolution bill to the committee on inland fisheries and game. They worried in their debate about the reputation of their state if such legislation should even be considered seriously. One legislator, Frederick Robie of Gorham, argued, "We should not go on record throughout the country as having entertained this bill even long enough to have wasted the State's money in having the thing printed."[18] Another Maine lawmaker worried that such a bill would only be supported by the most "unintelligent" and "illiterate" supporters.[19]

Among the batch of laws considered in 1927, many copied the language that proved so controversial in the Scopes trial. Many of the bills, as in the example of Alabama's first 1927 bill, stated their aim to "prohibit the teaching of the evolution theory in all the universities, normals, and all other public schools of Alabama."[20] Yet on closer examination, the bills debated in Alabama, Arkansas, South Carolina, and North Dakota all insisted on a favored educational status for Christian doctrine. Like the Tennessee law, the bills in these states demanded the prohibition of "any theory that denies the story of the Divine Creation of Man as taught in the Bible, and to teach instead that man has descended from a lower order of animals."[21] The North Dakota bill similarly banned "the teaching of any theory, denying the study of the Divine Creation of Man."[22] For many of the bills' supporters, such language merely banned evolution effectively. In effect, however, it revealed the underlying principle that the public schools must not challenge traditional interpretations of the Bible. It made the Edenic creation story the correct educational standard against which any teaching must be measured.

Further, a few of these 1927 bills made much more sweeping demands. West Virginia's Representative W. A. Street introduced a bill to ban "the teaching of any nefarious matter in our public schools." Florida's Representative Leo Stalnaker introduced a similarly expansive bill. His bill hoped to ban "the teaching as fact [of] any theory that denies the existence of God, that denies the divine creation of man, or to teach in any way atheism or infidelity, and to prohibit the use or adoption for use of any text book which [does so] . . . or that contains vulgar, obscene, or indecent matter." These proposed bills went far beyond the antievolution movement. They even reached beyond the fundamentalist movement to tap into a wellspring of traditional conservative sentiment about the nation's public schools. Nevertheless, their reception taught important lessons to fundamentalist school activists. No longer could demands for even broadly Christian public schools rely on an overwhelming majority of public support. The West Virginia bill met with little support, and although Representative Stalnaker's broad school bill originally received a favorable report from the Committee on Education and eventually passed the state House, it was eventually killed without a vote in the Florida State Senate. As we will see, fundamentalists learned a distressing lesson about the changing nature of mainstream American culture from these disappointing results.[23]

The results of such legislative activism also reveal a great deal about the educational goals of the fundamentalist movement in the later half of the 1920s. As the controversies over fundamentalist school bills continued nationwide, some state legislators hoped to pass more narrow antievolution bills. At the same time, however, some fundamentalists pressed for school bills with even broader language and even more frankly Protestant requirements for public schools. Of the thirty fundamentalist school bills, amendments, and resolutions introduced from 1926 until the end of the decade, seventeen demanded only the prohibition of evolution from public schools. The rest insisted more broadly on a special place for Protestant theology in public schools.[24]

In many of the state legislative battles, fundamentalist lawmakers doggedly refused to accept defeat. In Alabama, for instance, Representative C. O. Thompson leaped into action on the very first day the Alabama House of Representatives met after the Scopes trial. Thompson introduced a bill demanding respect for the traditional interpretation of the Genesis account of creation. Unfortunately for Thompson, his bill floundered in committee.[25] Undeterred, Representative D. G. W. Hollis, a Free Will Baptist minister from Pickens County,[26] proposed similar bills repeatedly in the summer of 1927. When opponents relegated his bills to parliamentary oblivion, Hollis maneuvered without success to earn his bill a hearing on the House floor.[27]

One state took a different path to a state law. In 1926 the Arkansas State Baptist Convention pressed legislators for a bill similar to the Butler Act in Tennessee. The next year, Baptist editor J. S. Compere and minister Ben M. Bogard successfully pressured representatives to approve the bill by a margin of fifty-one to forty-six. Unfortunately for fundamentalist supporters, the state Senate defeated the measure seventeen to fourteen. After this near miss, Bogard penned an initiative petition in favor of an educational antievolution statute. Arkansas

Baptists circulated the petition through their churches, and successfully put it on a referendum ballot in November of 1928. Arkansas voters overwhelmingly approved the measure. Arkansas' measure followed the language of Tennessee and Mississippi: teachers convicted of teaching evolution were subject to a stiff fine—between two hundred and one thousand dollars—and a loss of teaching credentials. With this vote, Arkansas became the last state in the 1920s to legally prohibit the teaching of evolutionary theory in public schools.[28]

In all these legislative battles, fundamentalists played an active role. Often, as in Arkansas, local fundamentalists took the lead. In many cases, though, national organizations continued their practice of organizing and speaking across states in which a legislative battle was taking place. In every case, however, new stereotypes about fundamentalism changed the ways these battles played out in the late 1920s.

One major difference after the trial was a spate of new self-proclaimed fundamentalist organizations. William Bell Riley, founder and leader of the World's Christian Fundamentals Association (WCFA), identified six "great movements" in America, all of which claimed to represent fundamentalism: The Fundamentalist League, The Defenders of Science vs. Speculation, the Bryan Bible League, the Anti-Evolution League of America, The Defenders of the Christian Faith, and the Bible Crusaders.[29] Some of these groups claimed a relatively long lineage. The Fundamentalist League, based in Los Angeles and headquartered at the Bible Institute of Los Angeles, had been active since at least 1920.[30] Most of them, however, had sprung up as a response to the events in Dayton. Some, such as the Bryan Bible League and the Defenders of Science vs. Speculation, made a defense of the fundamentalist position at the Dayton trial their explicit raison d'être. In spite of earnest efforts by their leaders, these two groups, like the Fundamentalist League, attracted relatively little national attention.

Others, such as the Defenders of the Christian Faith and the Bible Crusaders, made a claim to national leadership of the fundamentalist movement. These groups, along with the Supreme Kingdom, another upstart fundamentalist organization Riley could have included in his list, coalesced as explicit responses to the publicity of the Scopes trial. All three of these groups embraced the stereotype of fundamentalism bandied about by liberal commentators at the Scopes trial. In spite of the fact that these groups all quickly petered out as national organizations by the end of the decade, all three exercised liberal foes with militant language and extravagant promises of nationwide political activism.

Leading historians of twentieth-century evangelicalism have tended to disregard these start-up fundamentalist groups. George Marsden, for instance, dismissed the "furious activities" in the aftermath of the Scopes trial as a sign only of the movement's "bizarre developments."[31] Joel Carpenter only mentioned the Defenders of the Christian Faith in passing and he did not include either the Supreme Kingdom or the Bible Crusaders at all in his consideration of the fate of the 1920s fundamentalist movement.[32] These groups, though perhaps not as durable or intellectually coherent as other militant evangelical organizations in the 1920s, represented to themselves and to many contemporaries the essence of the fundamentalist movement. Fundamentalists hoping to emphasize the intellectual respectability of their movement may not have relished their company, but these post-Scopes activist groups embodied for many at the time the fundamentalist movement in the immediate aftermath of the famous trial.

The most active of these three new national organizations was the Bible Crusaders and Defenders of the Faith. This group had long been the dream of George F. Washburn. Washburn, a Boston-based real estate developer, had worked with Bryan in his Florida real estate speculations. An ardent fundamentalist, Washburn had long promised financial support to Bryan's evangelical endeavors.[33] After Bryan's death, Washburn rushed to fund a memorial university and to bankroll an organization that would continue Bryan's fundamentalist educational activism. From his adopted home in Clearwater, Florida, Washburn pledged to spend at least $200,000 "to prevent our becoming a pagan nation." The new organization promised a national fundamentalist magazine, the *Crusaders' Champion*, as well as a new radio station.[34] Privately, Washburn hoped to subsume all other fundamentalist organizations under his leadership.[35] Publicly, T. T. Martin, hired by Washburn as the field secretary for the new group, promised that the Bible Crusaders would pick up the "Flag of Fundamentalism" that Bryan had dropped upon his death in Dayton.[36]

The Bible Crusaders hoped to influence antievolution legislation nationwide by sending "Flying Squadrons"—modeled after temperance activists—to influence local decisions about fundamentalist-backed school laws.[37] Washburn invited local chambers of commerce to organize local meetings. The Crusaders would supply "scientists and lecturers of national reputation" wherever needed.[38] The Crusaders claimed a victory for this model in Mississippi in 1926. Washburn sent ten speakers, led by evangelist Martin, who together claimed to have spoken in three-quarters of the towns in that state in the run-up to the antievolution vote in the state legislature.[39]

Like the other upstart fundamentalist organizations, the Bible Crusaders explicitly embraced the stereotyped image of fundamentalism publicized by the Scopes trial. Many correspondents in Dayton from eastern urban newspapers assumed that fundamentalism was a regional affair. They repeatedly associated the movement with isolated Southern culture. The Bible Crusaders sought to capitalize on just this accusation. In their literature, they presumed a Southern base of support. They hoped to take advantage of the lingering lost cause mentality by promising supporters a "peaceful invasion of the North."[40]

The militant and paranoid language used by Washburn and other leading Bible Crusaders also matched nicely with popular stereotypes about the movement. Outsiders had long accused fundamentalists of anti-intellectualism, paranoia, and bigotry. The Bible Crusaders embodied these stereotypes. Their literature located the cause of cultural problems in the scheming designs of a small group. "Thirty years ago," Bible Crusader literature proclaimed, "five men met in Boston and formed a conspiracy which we believe to be of German origin, to secretly and persistently work to overthrow the fundamentals of the Christian religion in this country." According to the Crusaders, this conspiratorial group had claimed great success with their "deep, devilish, premeditated plan of propaganda." Only such a sustained conspiratorial campaign, they suggested, could make it appear as if mainstream culture in America had rejected the cultural hegemony of fundamentalist belief. In fact, the Crusaders claimed, 90 percent of Americans, including Catholics, were "fundamentalists at heart." Unfortunately, the small remaining minority had seized control of key cultural positions in education. This small minority would continue to pervert America's children unless the vast quiescent fundamentalist majority agreed to "rally to the flag of the Bible Crusaders, and stand like Spartans at Thermopylae."[41]

This kind of rhetoric in the years following the Scopes trial helped establish newly restricted definitions for fundamentalism. These fundamentalists presented themselves primarily as implacable and aggressive foes of modern America. Although many nonfundamentalists hoped that fundamentalists would retreat after the Scopes trial, other liberal enemies of the movement agreed with fundamentalist estimates of the movement's continuing strength. Maynard Shipley of the Science League of America wrote with alarm that "our country will ere long be converted into a relentless Fundamentalist theocracy."[42] In spite of Shipley's alarmist language, however, Washburn's group never achieved its ambitious goals. Not long after its founding, Washburn was forced by financial insolvency to pull out his backing of the organization. It had not achieved much of what it had hoped to do, but its Southern, anti-intellectual militancy helped to fix much more restrictive boundaries on the evolving definition of fundamentalism.[43]

The Defenders of the Christian Faith had a similar career in many respects. Like the Bible Crusaders, this group sprang to life in the aftermath of the Scopes trial. Like the Crusaders, the Defenders organized a group of itinerant speakers and activists who traveled extensively, promoting a fundamentalist agenda. They were also based in a region of strong fundamentalist popular support, and they tended to embrace the new stereotype of fundamentalism. In late 1926, the group began an energetic campaign. Unlike Washburn, at the end of a long career, the Kansas group was led by the young Gerald B. Winrod. At the time of the founding of the Defenders in November of 1926, Winrod was only twenty-six years old.[44] Winrod immediately began a campaign to ban evolution from Kansas schools. But he also had national ambitions. Like the Bible Crusaders, they sent groups of "Flying Defenders" or "Flying Fundamentalists" to attack evolution in venues around the country. They also published a magazine and hosted annual conferences. By the end of the decade, the Defenders claimed a membership of three thousand and a ten thousand-strong list of subscribers to their magazine. Winrod himself continued a national career as a conservative activist, but by the end of the 1920s, the Defenders had largely retreated from the national field to concentrate on fundamentalist school activism within Kansas.[45]

Nevertheless, their brief activism within the national fundamentalist movement helped to further cement the new image of fundamentalism among both fundamentalists and liberals. Like the Bible Crusaders, the Defenders explicitly organized as an effort to defend what opponents had attacked as fundamentalism. For one thing, they testily asserted their cultural victory in the Scopes trial. In 1927, for instance, the Defenders invited Riley to analyze the recent events in Dayton. "The trial was not a farce," Riley reassured his audience of Defenders, "but a tremendous event that has affected religious and educational life as nothing else in the last hundred years."[46] In the years following the trial, fundamentalists knew their movement had been mocked and attacked. They knew liberals and nonfundamentalists had concluded that fundamentalism was composed of bigoted rural thugs, and that their performance in the Scopes trial had been nothing more than low comedy. Gerald Winrod angrily refuted these attributions. Instead, he attacked his liberal opponents as the "most intolerant cult before the public today." Riley, Winrod, and other fundamentalist leaders reassured their readers and themselves that fundamentalism had continuing relevance and intellectual respectability.[47]

The Defenders also publicized their definition of post-Scopes fundamentalism by trumpeting their regional roots. They proudly hailed from Kansas, one of the regions secularist critic H. L. Mencken had dismissed as an "abyss of malignant imbecility."[48] And although the Defenders began their existence by proudly claiming their "progressive" past, they quickly shifted their rhetoric to match the new boundaries of fundamentalism.[49] Like the Crusaders, the Defenders embraced the stereotype of populist, paranoid anti-intellectualism. A year after the Scopes trial, the Defenders proclaimed their goal to "withstand the powerful, destructive, anti-Christian forces which threaten to annihilate revealed religion, blast away the foundations of civilization and introduce chaotic conditions . . . and save as many as possible from Satan's grip, who is working as an 'angel of light,' appearing in the form of so-called 'higher intelligence.'"[50] The Defenders' activism along these lines worked to confirm the liberal stereotype of fundamentalism. Although the Defenders themselves decreased their presence as a national organization after the 1920s, their acceptance and promotion of the new image of fundamentalism forced other fundamentalists nationwide to either embrace or reject its circumscribed boundaries.

The Supreme Kingdom experienced a similarly meteoric career. The organization was founded as an explicit response to the Scopes trial. It hoped to take advantage of the stereotyped image of fundamentalism in order to promote a powerful nationwide activist organization. Like the Defenders and the Crusaders, however, the big ambitions of the Supreme Kingdom were never realized. Nevertheless, like the other two groups, the limited activity of this group attracted a great deal of attention among both fundamentalists and their opponents. Its militant rhetoric and extravagant claims further intensified the pressure on fundamentalists to conform to the limited popular stereotype.

The Supreme Kingdom was founded in Atlanta in January 1926 by Edward Young Clarke. Clarke's involvement alone gave opponents fodder for attacks, for Clarke had previously served as a leader of the Ku Klux Klan. Accusations of vigilante violence and venality had dogged Clarke's reputation throughout the 1920s.[51] At its inception, the Supreme Kingdom declared lofty goals. The Supreme Kingdom would save a corrupt American culture by reforming its schools, religion, and politics. It planned to open offices in every major American city, and force "teachers, ministers, and officeholders" to submit to a hostile cross-examination by Supreme Kingdom functionaries. Clarke's announced goal was to "rebuild in the minds of our children the religion of our fathers."[52] A dispute between Clarke and New York fundamentalist leader John Roach Straton tarnished the image of both Straton and the Supreme Kingdom in the eyes of many fundamentalists. Clarke had claimed that Straton would be the group's well-paid itinerant revivalist. Straton, however, awkwardly distanced himself from his commitment. Straton claimed to have only agreed to join at an introductory level, not to take a leadership role.[53]

The Supreme Kingdom's image among liberals had been disastrous from the start. Due to Clarke's connection with the Ku Klux Klan, opponents dubbed the group yet another secretive group of violent reactionaries. Like the Bible Crusaders, the Supreme Kingdom's headquarters in Atlanta fulfilled stereotypes of fundamentalists as Southern holdouts. Shipley, the longtime antifundamentalist campaigner, quoted Clarke as threatening nationwide "bonfires" of "those damnable and detestable books on evolution."[54] In spite of Shipley's fears that the rise of the Supreme Kingdom after the Scopes Trial heralded "the overthrow of our present form of secular government," the group's activism never matched its aggressive self-consciously fundamentalist rhetoric.[55] It never managed to exert much influence outside of the Atlanta metropolitan area. Like other upstart fundamentalist groups, it hoped to capitalize on the publicity value of fundamentalism in the months and years following the Scopes trial. Also, like the other groups, it succeeded in confirming newly restricted definitions of fundamentalism, but it did not have the organizational ability to assert influence outside of a limited area for any extended period.

Fundamentalist antievolution campaigns had other leaders besides these new flash-in-the-pan organizations, however. More durable fundamentalist leaders also made school legislation one of their main goals in the years following the Scopes trial. As they had throughout the decade, they used evolution as a symbol of a cluster of ideas they found objectionable. For instance, when Minnesota fundamentalist leader Riley attacked the teaching of "evolution" at the University of Minnesota, he included a wide range of intellectual offenses in his indictment. Not only did students and professors promote the theory of evolution, Riley contended, but the teaching of many of those professors "closely approach[es] sovietism, and even anarchy." For Riley and other fundamentalist school activists, the campaign against "evolution" functioned as a shorthand for their desire to ban any idea that might challenge the supremacy of traditional Protestant intellectual presuppositions.[56]

Like Riley, fundamentalist activists traveled tirelessly across the nation, speaking to legislators and grassroots activists, debating with liberals and secularists, and organizing politically to pass school legislation. In some cases, as in the state laws passed in Mississippi and Arkansas, their efforts contributed to passage. In other states, such as North Carolina and Minnesota, years of dedicated and dogged activism and organizing by these fundamentalist leaders failed to secure passage of a statewide legal ban on evolution in schools, in spite of demonstrated popular support. While local factors often determined these struggles, the newly restricted boundaries of the fundamentalist movement also played an important part. Enemies used the stereotype of the Tennessee backcountry hillbilly to attack fundamentalism in general, and antievolution laws in particular. And fundamentalists often promoted these same stereotypes in their efforts to rally popular support for these measures.

The campaign for an antievolution law in Mississippi in 1926 was a case in point. In that state several fairly narrow antievolution bills had been languishing in parliamentary oblivion until Martin led a campaign to push a bill along.[57] Martin, now Director General of Campaigns for Washburn's Bible Crusaders and Defenders of the Faith, brought in ten lecturers to travel throughout the state. These speakers addressed local Baptist churches across Mississippi. They encouraged the congregations to send resolutions of support for the antievolution legislation to their state lawmakers. Martin himself traveled extensively, speaking in churches, rented auditoriums and movie theaters.[58] As the debate reached the floor of the state House of Representatives, Martin received an invitation to address the lawmakers themselves. The tireless drive of Martin and the Bible Crusaders worked. In March 1926, the Mississippi legislature passed the nation's second statewide law banning the teaching of evolution outright.[59] One contemporary writer estimated that Bible Crusaders lecturers spoke in at least three-fourths of Mississippi's cities.[60] Another called the successful passage of the bill a "remarkable achievement of Fundamentalism."[61] As one state senator admitted, "The arguments of these gentlemen [the Bible Crusaders] were forceful and convincing, and went a far way in changing sentiment for the bill."[62]

Fundamentalist success in Mississippi was due in large part to the publicity of the Scopes trial the previous summer. After all, Martin would never have been able to bring such pressure to bear in Mississippi were it not for the generous financial backing of George Washburn, and Washburn had been immediately motivated by the trial to bankroll a national fundamentalist organization.[63] Martin's activism in Mississippi was the first produce of Washburn's investment. Fundamentalist speakers were also able to exploit the new stereotype of fundamentalism as an efflorescence of Southern exceptionalism. National and northern newspapers often presumed that antievolution sentiment was a uniquely Southern phenomenon. Fundamentalists took advantage of that stereotype in order to pressure Mississippi lawmakers to vote for the antievolution law as a test of their Southern legitimacy. Mississippi lawmakers clearly felt this popular pressure. Although the bill had received adverse reports from both House and Senate committees, the lawmakers on the floor could not ignore the galleries packed with grassroots antievolution activists. Furthermore, the fundamentalist campaign had deluged representatives and senators with petitions from Baptist congregations across the state. In the face of such adamant public support, Chancellor Alfred Hume of the University of Mississippi limited his opposition to the bill to a private letter to the governor. Supporters of the bill overwhelmed such tepid opposition.[64]

In an address to the House of Representatives, Martin manipulated the new stereotype of fundamentalism to rally political support. He challenged legislators to see the two sides in the issue as two distinct cultural archetypes. By opposing the law, Martin suggested lawmakers would be pandering to the "fulsome praise of a paganized press." He acknowledged that "the law will bring on Mississippi the ridicule and abuse from the North that have been heaped upon Tennessee." But Martin belittled the censure that had come to surround the fundamentalist movement. Martin assured Mississippi lawmakers that the "ridicule and scorn and abuse" that opponents might attach to Mississippi's proposed law mattered only to those who cared about the opinion of "Bolshevists and Anarchists and Atheists and Agnostics and their co-workers." Supporters of the law, however, had the chance to protect "the faith of the children of Mississippi in God's Word and in the Savior."[65] In the wake of the Scopes trial, only a minority of Mississippi politicians felt they could safely vote against the new stereotyped fundamentalism. The Mississippi House approved the Evans Bill by a vote of seventy-six to thirty-one, the Senate by a margin of twenty-seven to sixteen.[66] Martin had recognized and successfully exploited the publicity of the Scopes trial to paint the political choice in such sharply defined terms. Lawmakers could side with either the mainstream press or the new stereotype of Southern populist anti-intellectual fundamentalism. In Martin's argument, fundamentalism had become an acknowledged object of ridicule among mainstream American opinion. Such ridicule, however, could serve as a sign of authenticity among some audiences.

Similar tactics proved harder to implement elsewhere. After his success in Mississippi, Martin hoped to bring his campaign to North Carolina. Martin had high hopes for success. After all, the campaign in North Carolina had had a long history, and had achieved a great deal of popular success. Martin himself had initiated public debate about evolution in North Carolina with a series of articles in 1920 attacking the administration of Baptist Wake Forest College. The state legislature had considered antievolution bills in early 1925, and fundamentalists had applied themselves unsuccessfully to their passage.

In addition to Martin, Straton, Riley, Bryan, and the popular "baseball" evangelist Billy Sunday had toured the state. Lesser-known evangelists pushed just as energetically for the fundamentalist-backed school legislation. North Carolina native A. C. Dixon wrote and spoke throughout the state, as did Baxter "Cyclone Mack" McClendon, a close associate of the "Texas Cyclone" J. Frank Norris. In spite of this fundamentalist blitz, the Poole Bill, as the 1925 North Carolina legislation was known, was narrowly defeated.[67]

Martin hoped to reverse that decision in 1926. He brought his Bible Crusaders cadres to North Carolina, fresh from their victory in Mississippi. To further focus fundamentalist attention on the North Carolina campaign, Riley hosted the annual conference of the national Anti-Evolution League and the Bible Crusaders in Charlotte in May. At the conference, Martin diplomatically flattered local fundamentalist leaders by assuring them that their activism in North Carolina would soon determine the results of the fundamentalist school campaign nationwide. Soon thereafter, Martin and his assistants set up four district headquarters and began planning debates, distributing literature, and organizing local affiliates of Martin's Anti-Evolution League.[68] Riley's World Christian Fundamentals Association sent the accomplished antievolution dialectician Arthur I. Brown to help Martin's "thirty day whirlwind campaign."[69] Brown promoted himself as a leading fundamentalist scientist, and even many nonfundamentalist North Carolinians accepted his claim to be the "greatest scientist in all the world."[70] With such expert assistance, a committed statewide organization, and strong popular support, Martin and other fundamentalists expected a decisive victory in North Carolina.

Unfortunately for Martin, the new image of fundamentalism worked against him in North Carolina. Liberals in North Carolina used the new stereotype to tar supporters of the antievolution bill as ignorant bigots. Furthermore, fundamentalists in North Carolina split over the new meanings of their movement. They fought among themselves and with national leaders. A majority of locals agreed that their definition of fundamentalism precluded assistance from national leaders. In their opinion, fundamentalism had come to demand a primarily local, anticosmopolitan campaign.

One of the liberal leaders in the Tar Heel State was Harry Chase, president of the University of North Carolina. Like other university presidents, Chase argued that an antievolution law would drive scholars and scholarship out of the state. In the aftermath of the Scopes trial, Chase manipulated the popular new image of fundamentalism. Echoing the accusations of Clarence Darrow and Mencken, Chase accused supporters of the Poole Bill of leading North Carolina into "intolerance, bigotry, and fanaticism."[71] Other critics made the association to the events in Dayton more explicit. The *Charlotte Observer*, usually friendly to the antievolution crusade, opined that the movement had become "a cheap show of the common order." In the opinion of *Observer* editors, North Carolina had seen "enough of this monkey business for quite a spell."[72]

More daunting to Martin and the other professional fundamentalist activists was opposition from local fundamentalists. A local group had organized in early 1926, calling itself first the Committee of One Hundred, representing all one hundred counties. It soon changed its name to the North Carolina Bible League. This group proudly identified itself as a group of "snorting fundamentalists," but they fought with Martin and with one another about the proper strategy for their school campaign.[73] The Committee refused to invite Martin to their meetings, nor would they grant him space to hold a debate. According to historian Willard Gatewood, local fundamentalists feared Martin's reputation from the Scopes trial would link their antievolution campaign with the anti-intellectualism and antiurbanism associated with the Tennessee trial.[74] In addition, they knew that any association with such outsiders would open them to attacks of letting in "unwanted foreigners."[75] At one turbulent meeting, the committee resolved that "no outside help is wanted" in the drive to bring North Carolina's schools into line with fundamentalist belief.[76]

Worse, the North Carolina committee fought among itself. Members could not agree on the meanings of fundamentalism, and "straight-laced" fundamentalists bitterly—and almost violently—fought with "liberal Fundamentalists" over the nature of fundamentalism in the wake of the Scopes trial. Leaders of the "liberal" faction were eager to distance themselves from the new image of fundamentalism. Liberal leaders such as A. A. McGeachy, pastor of the Second Presbyterian Church of Charlotte, and W. E. Price, a Presbyterian elder from Mecklenburg County, denounced the "lack of tolerance and un-Christlike spirit" of the other fundamentalists. These two former fundamentalist leaders abandoned the movement due to the newly restrictive boundaries it developed after the Scopes trial, and they correctly predicted that many more fundamentalists would soon abandon fundamentalism for similar reasons.[77]

Even faced with such disarray over the meanings of fundamentalism, the local fundamentalists and Martin's itinerants managed to send petitions with fifteen thousand names to the state legislature. However, the wounds of the campaign took their toll. After three hours of hearings, the House Committee on Education effectively killed the bill. In North Carolina, the contested boundaries of fundamentalism played a key role in this failure to pass a law. Liberal opponents used the stereotypes of the Scopes trial as an effective way to dismiss supporters of the bill as bigots and clowns. Confused definitional boundaries also caused fundamentalists to fight over the proper definition of their movement. As would many 1920s-era fundamentalists nationwide, many fundamentalists in North Carolina did not accept the newly restricted bound-aries on fundamentalism. In the wake of the Scopes trial, these fundamental-ists agreed with liberals that fundamentalism had come to imply intolerance and bigoted aggression. These leaders quietly abandoned fundamentalism, even though they still supported traditional Protestant religiosity in North Carolina's public schools. Nor were local grassroots activists able to combine local organiz-ing networks with the resources of professional activists, as they had in Missis-sippi. Perhaps because Martin himself was a native Mississippian, he had been able to work with local fundamentalists much more effectively in that state. In North Carolina, however, Martin's intense involvement became a liability rather than an asset.[78]

Similar frustrations dogged the fundamentalist drive for an antievolution law in Minnesota. The campaign had always been an uphill battle in that state, but Riley was determined to pass a law in his adopted home state. Riley had first attached himself to the antievolution movement in October 1922, after hearing Bryan deliver one of his rousing antievolution speeches in Min-neapolis. That very week, Riley had organized Twin Cities pastors into an antievolution coalition. Soon, Riley officially turned his ad hoc group into the Anti-Evolution League of Minnesota. For the next several years, as Riley led the fundamentalist campaign nationwide for antievolution laws, he also maintained a desultory crusade in Minnesota. Riley debated with prominent local evolution supporters, published antievolution articles in local papers, and made speeches across the state.[79]

In 1927, Riley decided the time was ripe for Minnesota's own law. He wrote and introduced a bill into the state legislature, then applied the same pressure in his home state that he had across the country. Riley assiduously flattered and courted state legislators and local bigwigs. He brought in several lecturers, including fundamentalist scientists Arthur I. Brown and Harry Rimmer. Gerald Winrod also campaigned energetically around the state. Riley himself delivered sixty-five speeches in support of the bill and the World's Christian Fundamen-tals Association speakers lectured, preached, and debated evolution supporters in two hundred Minnesota towns altogether.[80] The World's Christian Funda-mentals Association also produced and distributed a pamphlet for those who needed more evidence.[81] The pamphlet exposed evolutionary teaching in text-books used in Minnesota public schools and attacked evolutionary and atheistic teaching at the state university. To drive its point home, Riley even quoted arch-foe Clarence Darrow. In 1924, Darrow had defended accused murderer

"Babe Leopold" on the grounds that Leopold had been taught amoralism at college. "The University that taught it would be more to blame than [Leopold] is," Darrow had asserted. According to Riley's logic, it followed that if the state colleges of Minnesota continued to teach evolution, the next generation would develop into amoral monsters.[82] In spite of this intense campaign, the state Senate defeated the antievolution bill soundly, by a margin of fifty-five to seven.[83] As usual, Riley obstinately claimed victory. Although the drive to ban evolution "lost in the legislature," Riley claimed, it "has been won in the State."[84]

The failure of the antievolution bill in Minnesota puzzled Riley, with good reason. As they had in North Carolina, fundamentalists from across the United States and Canada had traveled tirelessly throughout the state to promote the bill. The bill also had significant popular support. When President Lotus Coffman of the University of Minnesota asked fellow Minnesota college and university presidents for help opposing the law, very few volunteered. The fundamentalist school bill had too much popular support; few college or university presidents besides Coffman dared side against it publicly. Yet in spite of this recipe for success, the bill was soundly defeated. As in Mississippi and North Carolina, new stereotypes of the fundamentalist movement played a key role. Opponents were able to discredit the bill as, at best, an ignorant joke. Potential supporters, especially the numerous conservative Lutheran population in Minnesota, were scared off by the new connotations of the fundamentalist movement.[85]

Since the Dayton trial, a wider scope of Minnesota's journalists and mainstream intellectuals assumed fundamentalism represented the forces of aggression and ignorance. Alvah Eastman, a writer for a St. Cloud newspaper, attacked Riley's "vicious . . . bitter, hate-making" campaign. According to Eastman, such a controversy could never influence "thinking people." The publicity of Darrow's charges at the Scopes trial had given increased weight to such charges among liberals. David Swenson, a philosophy professor at the University of Minnesota, turned up his nose at the challenge to debate fundamentalist intellectuals. According to historian Ferenc Szasz, Swenson "scoffed at Harry Rimmer's academic credentials and claimed he was not qualified for intelligent scientific discussion." Instead of a debate among philosophers, scientists, and theologians, the newly restricted image of fundamentalism made this a cultural clash between two incompatible value systems. Like Swenson, many Minnesota liberals did not recognize the legitimacy of fundamentalist credentials. To them, the fundamentalist educational movement had become a sad and possibly dangerous joke.[86]

Even some fervent antievolutionists in Minnesota refused to support the law. Many conservative Lutherans in Minnesota opposed the teaching of evolution and all that went with it, but they were more committed to preserving an inviolate division between religion and government. Riley worked hard to counter this belief. He assured his Lutheran audiences that the proposed law "does not involve a union of church and state."[87] After all, Riley's Baptist tradition had also long fought to keep the two separate. Yet most Minnesota Baptists supported the law, while Minnesota Lutherans largely did not. Riley's personal leadership must have made part of the difference. Baptists could look to Riley and other leading Baptists at the head of the antievolution and fundamentalist movements. Lutherans had fewer fundamentalist role models to look to. In addition, unlike in Mississippi or North Carolina, voters in Minnesota could not be compelled by the new regional stereotype. Since the Scopes trial, opposition to evolutionary theory had come to appear in the eyes of many as a peculiarity of isolated Southern communities. For Mississippians, this became a badge of honor. But northern voters felt no such regional identification with the new stereotype. Furthermore, Lutherans in Minnesota tended to have a much more recent immigrant history than local Baptists.[88] After the publicity of the Scopes trial, public identification of fundamentalism with such groups as the anti-immigrant Ku Klux Klan and its offshoot the Supreme Kingdom must have scared some Lutherans away from the movement. As Riley recognized, these conservative Minnesota Lutherans continued to oppose the teaching of evolution, but they did not wish to be associated with the new public face of fundamentalism.[89]

Riley did not despair after the loss in his home state. He pushed for antievolution laws nationwide and learned from his defeat. Although Arkansas' legislature defeated an antievolution bill, Riley helped local fundamentalists in that state take their case directly to the voters. Local fundamentalist leaders including Baptists Ben Bogard and J. S. Compere had struggled throughout the decade to bring their antievolution bill to a vote in the state Senate. Since the early 1920s, Bogard had invited national leaders Bryan, Martin, and Riley to Arkansas to help convince local congregations to send resolutions of support to their state legislators. In spite of such coordinated efforts, the state Senate effectively killed the bill in 1927 through parliamentary maneuver. Compere and Bogard, with the assistance of Martin and Riley, formed the American Anti-Evolution Association for Arkansas in response. Together, these activists garnered enough signatures to put a narrowly worded antievolution initiative on the November 1928 ballot. Arkansas voters overwhelmingly approved the measure, 108,991 to 63,406.[90]

This success owed a great deal to the lessons of Mississippi, North Carolina, and Minnesota. As they had in Mississippi, fundamentalists in Arkansas were able to coordinate long-term local campaigns with intense campaigns by well-known national leaders. They used established networks of local fundamentalist congregations to communicate and organize petitions. They did not fight among themselves as they had in North Carolina, and they faced no liberal opposition as determined and capable as that of university presidents Lotus Coffman in Minnesota or Harry Chase in North Carolina.

Also, in Arkansas, Riley won at least one battle over the political meanings of fundamentalism. As the controversy heated up, Riley debated the issue with prominent atheist Charles Smith. In the debate, Smith maladroitly used traditional liberal arguments about the definition of the fundamentalist movement. Smith accused supporters of the law of pandering to "rural ignorance."[91] Earlier, Smith had made public statements calling Arkansas a "joke," and laughing that "fundamentalist hill-billies dominate the state."[92] Such undiplomatic attacks gave Riley an opportunity to defend the maligned stereotype of fundamentalism. "'RURAL' intelligence," Riley countered, "and urban intelligence and metropolitan intelligence will put this infamous theory where it belongs, on the day of the November election." But Riley did not limit his appeal to this defense of fundamentalism. He also offered a positive identity with which voters could identify. "If you want [your children] to be Christian, clean and wholesome, upright, sane, sensible, self-respecting, keen to exhibit brotherly love and worship the true God," Riley exclaimed, "vote for the Bill." Arkansas voters agreed.[93]

In each of these states—Arkansas, Minnesota, North Carolina, and Mississippi—there was no simple reason for the success or failure of antievolution legislation. In an effort to understand why some state legislatures passed such laws, while others did not, political scientist Michael Lienesch has recently offered four critical elements for success. Following the description of political scientist Sidney Tarrow, Lienesch identified four necessary factors for successful political agitation: "access," or the relative ease activists have in approaching levers of political power; "alignments," or the "stability or instability of elite interactions"; "availability of allies" among political elites; and "cleavages or divisions among elites." Lienesch argued that such elements were often present in southern states during the 1920s. In Oklahoma, for instance, where state legislators approved an antievolution resolution in 1923, antievolution activists enjoyed easy access to power in the state legislature. Throughout the South, Lienesch pointed out, elites scrambled to find allies among those outside the halls of power. Fundamentalists and other antievolution activists happily took advantage of these divisions and alignments. Although Lienesch did not make the point, the presence or absence of powerful allies also explains some of the legislative failures and successes. For instance, in Minnesota, the 1927 antievolution bill failed miserably in spite of widespread popularity and in spite of energetic fundamentalist political activism. Perhaps the failure of Riley to line up the support of conservative, antievolution Minnesota Lutherans as political allies explains his resounding defeat. In contrast, Martin's success in Mississippi was due, in large part, to his successful appeal to fellow Mississippi Baptists to enroll as allies in the fight for a new state law. Similarly, the activism of Baptist congregational networks in Arkansas played a crucial role in the passage of that state's law.[94]

However, as Lienesch acknowledged, his four-point outline does not satisfactorily account for every success or failure of such laws during the 1920s. For instance, Lienesch described the complex legislative climate in North Carolina. North Carolinians had abundant "opportunity," to use another of Lienesch's terms, to pass an antievolution law. They met all four of Tarrow's elements for success, but antievolution activists never succeeded. In spite of the abundant access to power of antievolution activists, the instability of political elites, the depth of allies for the antievolution movement, and divisions among elites at least as powerful as in other states, North Carolina never passed such a law. Local factors, such as the energy and charisma of University of North Carolina president Harry Chase, determined the outcome of the political battle, in spite of the great potential for success according to Tarrow's four points.[95]

In many cases, such exceptions become glaring enough to make any simple analysis of the reasons for success or failure of antievolution laws frustratingly inexact. Fundamentalists, after all, had built extensive and well-organized support networks in Minnesota and North Carolina as well as in Mississippi and Arkansas, yet they could not use those networks to pass a statewide antievolution law. University of Minnesota president Lotus Coffman had failed in Minnesota to organize a powerful university lobby against the antievolution bill, yet the Minnesota bill was soundly defeated.

One common factor in each of these states was the political use of new stereotypes of fundamentalism. Activists on both sides worked to manipulate those new stereotypes. In some cases, fundamentalists hoped to rally support around an image of fundamentalism as militantly conservative Southern pride. Liberals sought to discredit fundamentalist school laws as merely the result of ignorant backwoods fanaticism. This complicated contest demonstrates the difficulty in assigning simple reasons for passage or failure of these state laws. Every state experienced different controversies. In North Carolina, liberal activists successfully attacked the 1926 antievolution bill as a ridiculous, anti-intellectual, nonnative import. Minnesota voters, for their part, were unmoved by fundamentalist appeals to southern pride. In Arkansas, in contrast, voters identified attacks on the antievolution referendum with the smug insults of New York atheist Charles Smith. In Mississippi, only seven months after the dramatic Scopes trial, Mississippi lawmakers were pressured to side with their Southern identity, now linked irresistibly to the fundamentalist movement.

In practice, liberals and fundamentalists alike worked to promote these new identities, even when they did not intend to. When famous secularists such as Mencken attacked fundamentalists as southern imbeciles, many southerners reacted by clinging defiantly to the new image of fundamentalism. Conversely, when Edward Young Clarke, a famous Klan leader, started a new fundamentalist organization, many conservative Catholics and Lutherans hastened to separate themselves more distinctly from the movement. In both cases, local political realities trumped fundamentalist efforts to pass antievolution laws. In spite of energetic and efficient political organizing in both Minnesota and North Carolina, contested meanings of fundamentalism allowed enough lawmakers to oppose antievolution bills safely. In Arkansas and Mississippi, however, local fundamentalists combined with professional activists to successfully deliver enough votes to pass strict new school laws.

Both losses and victories in this so-called antievolution legislative campaign shaped fundamentalist thinking about the nature of their role vis-à-vis the nation's public schools. Many fundamentalists who had joined the movement with grandiose hopes of quickly reestablishing cultural control of public education learned a bitter lesson. After the Scopes trial, they found themselves encumbered by the new public image of their movement. Although they occasionally used it to their advantage, many fundamentalist school activists smarted at the way their school laws could not shake association with the famous "monkey" trial. Similarly, although many opponents of the fundamentalist school laws happily used the stereotypes of ignorant rural backwoodsmen to discredit the intellectual credibility of their fundamentalist foes, they also found to their chagrin that fundamentalists could appeal to positive versions of those same stereotypes to rally support.

CHAPTER 7

College Controversies after Scopes

State legislatures were not the only forums in which these issues played out in the later years of the 1920s. College and university campuses continued to roil with controversy over issues of evolution, fundamentalism, and atheism. Fundamentalists and liberals fought over issues of control and academic freedom at both public schools and denominational colleges. Although these battles took place across the country, the controversies in North Carolina and Minnesota attracted the most sustained attention and optimism from fundamentalist activists.

As they had with the legislative campaign, liberals attacked fundamentalists as ignorant, aggressive hillbillies. As we have seen, some fundamentalist leaders embraced and promoted the image of fundamentalism as Southern, paranoid, and anti-intellectual. But many more fundamentalists struggled to come to terms with these powerful new stereotypes of fundamentalism as they fought for control of higher education. In many cases, fundamentalists grudgingly accepted a new cultural reality; they recognized that regaining cultural control of the major public universities would be impossible. In some schools, evangelical students articulated a new identity as beleaguered minorities in the secularizing world of higher education. Other fundamentalists established fundamentalist schools of their own. Some of these new schools succeeded beyond their founders' expectations. Often, however, ambitious fundamentalist college founders met with disappointing defeat. Just as in legislative battles, new arguments about the proper definition of fundamentalism played an important role in these struggles.

As it had with fundamentalist-backed school legislation, North Carolina became a hotbed of fundamentalist activism for control of higher education. And, just as with the legislative campaign, fundamentalists fought for much more than the prohibition of evolutionary teaching. They hoped to limit or prohibit any teaching that threatened to weaken the faith of evangelical college students. However, similarly to the legislative campaign, "evolution" functioned as both a symbol and an example of the kinds of teaching and thinking that fundamentalists found pernicious.

By 1925, the public colleges of North Carolina had acquired an image as proevolution, antifundamentalist institutions. One incident that had fired fundamentalist sentiment against the system was a pair of articles that had appeared in a journal based at the University of North Carolina. The articles appeared in the scholarly *Journal of Social Forces*, edited by the well-known sociologist Howard Odum. The first article, "The Development of the Concept of Progress," by L. L. Bernard of Cornell University, analyzed various myths, including the foundational stories of Christianity. Bernard concluded that religion, including Christianity, expressed merely "the projection of [early humanity's] hopes and the personification of his ideals . . . gods . . . whom he had created and he thought had created him."[1] The second article, "Sociology and Ethics: A Genetic View of the Theory of Conduct," by Harry E. Barnes of Smith College, contained similarly inflammatory conclusions. Barnes condemned religion as "superstition and accident, elaborated into beliefs and conviction." Instead of a divine revelation, the Bible was merely "the product of the folkways and mores of the primitive Hebrews . . . and the personal views of religious reformers of all grades from Jesus to Paul."[2] Both Barnes's and Bernard's articles articulated arguments about the roots of Christianity that had long fueled fundamentalist anxiety. Due in part to energetic circulation of the two articles by fundamentalist activists, fundamentalists across North Carolina and the nation loudly protested the state university sponsorship of such publications. Baptist and Presbyterian organizations across North Carolina lodged formal protests against the state university for sponsoring such offensive analyses.[3]

Fundamentalist sentiment was further outraged in 1925 when Albert S. Keister, a professor at the North Carolina College for Women in Greensboro, informed a class full of public school teachers that the theory of evolution was a gift of science, since it forced people to overcome their belief in Genesis, which he called "a form of mythology." Fundamentalists quickly turned Keister's comments into a local cause célèbre.[4]

Fundamentalists and other conservatives also smarted at the strong leadership of Harry W. Chase, president of the University of North Carolina. Chase boldly led a movement to oppose the Poole Bill, which sought to criminalize the teaching of evolution in North Carolina's public schools. Chase publicly worried that passage of such a bill would antagonize top faculty. In the end, Chase successfully defeated the bill by arguing for the need for freedom of thought among faculty. Chase's winning argument defined the debate as a question of intellectual freedom versus ignorant bigotry.[5]

Such an image, however, also energized fundamentalist activists in their fight for control of higher education in North Carolina. Local fundamentalists in the Committee of One Hundred had not succeeded in passing a statewide legal ban on the teaching of ideas that threatened evangelical belief, but they did succeed in pressuring scholars and administrators at the public institutions of higher education to consider fundamentalist sentiment. After Professor Keister's alleged comments raised a storm of publicity, fundamentalists mobilized to oust Keister from his professorship. While the administration of President Julius Foust succeeded in maintaining Keister, it was a very close vote among the board of trustees. In order to secure his employment, Keister was forced to agree to avoid any controversial statements in his future teaching. President Chase of the flagship University of North Carolina in Chapel Hill made similar political concessions. He postponed the publication of a book on evolution by a member of the university faculty. He also worked assiduously to promote an image of a university "loyal to both religion and science."[6]

Chase and Foust were right to be careful. They could not assume their legislative victory had saved them from political pressure. While the atmosphere in the aftermath of the Scopes trial had given them an advantage in the fight against statewide fundamentalist school legislation, it had not eliminated fundamentalist interest in higher education. But it did make fundamentalists more concerned about the public image of their movement. As one local observer noted, "Because we have the reputation of being the most progressive of the Southern states; . . . because public opinion generally is more enlightened and fair-minded than that in Tennessee . . . even our convinced Fundamentalists hesitate to make this state also a laughingstock."[7] Antievolutionists in North Carolina might not have been able to secure an outright legal ban on the teaching of evolution, due to the political power of the new stereotypes about fundamentalism. However, those same activists could be counted on to apply intense pressure to the state's public colleges and universities. The publicity of the Scopes trial had not implicated those higher educational campaigns as fiercely as they had the so-called antievolution battle. Leaders such as Chase and Foust knew they needed to tread carefully in order to fight off fundamentalist challenges to their public universities.

Denominational colleges in North Carolina experienced an even more caustic fight. The most prominent conflict occurred at the most prestigious Baptist school in the region, Wake Forest College. After the Scopes trial, Baptists persisted in their criticism of William Louis Poteat. Poteat had successfully defended himself and his administration in 1922, but fundamentalists attacked his policies unrelentingly. National leaders such as T. T. Martin and J. Frank Norris had long publicized claims by Baptist laymen that Wake Forest had turned their children into evolutionists and atheists.[8] C. A. Jenkins, pastor of Zebulon Baptist Church in Zebulon, North Carolina, inspired his congregation to pass a resolution in 1923 condemning Poteat's Wake Forest, requesting the dismissal of any faculty member in Baptist schools who taught or held "the theory of evolution in any of its forms."[9] Local fundamentalists redoubled their assault in the summer and fall of 1925, in the immediate aftermath of the Scopes trial. They accused Poteat of mismanagement and of neglecting popular Baptist antievolutionism. County Baptist associations publicly condemned Wake Forest's accommodations with evolutionary theory and materialism. One association even suggested a new funding system by which local churches could fund Baptist projects while withholding funding for the college.[10]

President Poteat defended himself vigorously. He repeatedly used the public reputation of the Scopes trial to attack his fundamentalist opponents. His son, also a teacher at Wake Forest, had earlier accused prominent fundamentalists such as Norris, William Jennings Bryan, and New York fundamentalist pundit John Roach Straton of being "bigots."[11] President Poteat himself had gone on the offensive with a lecture series in early 1925, in which he called fundamentalists "misguided men"[12] whose wrongheaded political activity "comes in the wrong century."[13] Poteat asserted, "About the principle and fact of evolution there is no question."[14] These accusations rallied liberal supporters, but they also reinvigorated the fundamentalist opposition. Poteat was forced to fight to keep his job in 1925, in the face of widespread calls for dismissal. He managed to do so, but just barely. At the state Baptist convention in late 1925, Poteat and his supporters again used the new stereotyped identity of fundamentalism as a weapon with which to chastise their fundamentalist opponents. Liberal Baptists blasted North Carolina fundamentalists as "violent," "ignorant," "misguided," "intemperate and bitter."[15] President Poteat himself manipulated the new popularity of this fundamentalist image by declaring that his liberal Baptist policies promised the only path to "respectability." The public image of fundamentalism after the summer of 1925 had come to imply anti-intellectualism, demagogy, and rural isolation. Poteat recognized this strategic advantage and promised North Carolina Baptists that only aggressively liberal religion would allow them to avoid this image.

Fundamentalists around the nation watched events in North Carolina closely. Even after some local fundamentalists had distanced themselves from national organizations in the wake of the Scopes trial, William Bell Riley, for one, devoured news coverage of the fights for an antievolution law and for control of higher education.[16] In March 1926, inspired in part by the ardent localism of North Carolina's Committee of One Hundred, Riley sought an invitation to speak closer to home, on the University of Minnesota campus, about banning evolution from public colleges. A nervous dean, however, revoked Riley's invitation at the last minute, and an outraged Riley spoke to a crowd of 5,500 at a private venue instead.[17]

In his speech, Riley wrestled with the new public perception of the fundamentalist movement. On one hand, he disputed the popular idea that fundamentalism represented bigotry. On the other, he played to the populist anti-intellectualism of the new image. Riley blasted the public university along familiar lines, listing textbooks that promoted atheism or evolution, and riling up the crowd by asking, "Do you want that kind of teaching in your state university?" to which the crowd chanted, "No! No!"[18] He ended his speech with a populist call to oppose the claims of "a dozen regents or a hundred Darwinized or Germanized, deceived and faithless professors." Riley claimed, in the end, to be speaking for "the God-believing, God-fearing Minnesota majority."[19] But he did not limit his polemic to these populist, majoritarian, anti-intellectual arguments. Riley also accused university administrators of shortsighted prejudice. Modernist Protestants, Riley pointed out, had been invited to speak on campus. Even when Riley had offered to pay the honorarium for fundamentalist speakers, however, the university refused. The university, Riley contended, denied students' "right to hear two sides of a controverted subject." Riley condemned the university in terms made popular at the Scopes trial. In this case, however, Riley attempted to use them against liberals. University policy, according to Riley, represented a "contemptible piece of prejudice in theology . . . unbecoming an institution that belongs to the whole people of the state." Riley charged the university leaders with refusing to allow students to think for themselves and accused them of intimidating students into accepting evolutionary theory.[20]

Cowed by the popularity of Riley's appeals, the administration of the University of Minnesota relented and invited Riley to give a series of talks on campus during the 1926–27 school year. An excited Riley used his rostrum to blast the university. He charged professors with actively undermining students' faith by "propagating" evolution. It was not only evolution that angered Riley. He accused professors at the university of promoting socialism and other political ideas that Riley found appalling. Even student leaders, according to Riley, were so "steeped" in evolution that they were leading other students astray.[21]

In his speeches, Riley carefully cultivated an image of fundamentalism as polite, cultured, and scholarly. When a student released a live monkey on stage during one of his speeches, Riley urbanely laughed it off. He also condescendingly complimented his adversary, University of Minnesota president Lotus Coffman, for Coffman's polite reception of Riley's ideas. Riley encouraged other fundamentalists across the country to make similar speeches at local colleges and universities, but he carefully insisted that only "competent" men make such speeches. Riley worried that demagogic anti-intellectual diatribes on college campuses would only drive the fundamentalist movement further into the stereotype promulgated at the Scopes trial.[22] Riley's presentation scored some successes. The university paper complimented Riley as "A Very Nice Looking Man," but it pointed out that most students still did not accept his views nor did leaders at the public universities in Minnesota. President Coffman continued his public role as the unofficial state spokesperson against antievolution legislation. In the end, Riley's continuing advocacy of stricter rules against evolutionary teaching in public higher education did not make any significant difference in the teaching at those schools.[23]

Riley had more success in his renewed attacks on nearby Carleton College. In 1926, inspired by the victories in Tennessee and at Baylor College in Texas, he introduced a resolution at the Minnesota Baptist Convention to cut off all funding from this school. Riley charged that Carleton was "not only saturated with Modernism, but is completely committed to Rationalism vs. Revelation." According to Riley's proposed resolution, Carleton was not only unorthodox but even "rankly liberal, with a tendency to Unitarianism."[24] This resolution languished in committee for two years, but it finally passed in October 1928 by a vote of 172 to 135.[25] Carleton was "disfellowshipped," cut off from any affiliation with the Minnesota Baptists.[26] In the long run, this rejection did not inflict any serious injury on Carleton's finances or reputation, and it soon became a respected center of antifundamentalist liberalism in the upper Midwest.[27]

The campaigns at Wake Forest and Carleton received a good deal of publicity due to the notoriety of the fundamentalists involved and the reputations of the colleges. However, in the years following the Scopes trial, similar battles raged at lesser-known denominational schools across the United States and Canada. Inspired by the well-publicized attempts of professional activists to battle for control of college curricula, local fundamentalists investigated their own denominational schools. At Baptist Ouachita College in Arkadelphia, for example, the Arkansas Baptist convention fired the president for refusing to sign a fundamentalist confession of faith.[28] Ohio fundamentalist activists also harassed the administration of the (Presbyterian) College of Wooster in Ohio. Fundamentalists demanded the firing of suspect teachers. As with all of these collegiate controversies, the objectionable teachings reached beyond evolution. The Wooster fundamentalists pressured its faculty to sign a creed affirming "belief in the 'Inspiration of the Bible,' 'the Virgin Birth,' 'the Atonement,' 'the Bodily Resurrection,' and 'the Miracles of Jesus.'"[29] Local fundamentalists made similarly broad demands of the administration at the (Baptist) Kalamazoo College, in Michigan. Activists attempted unsuccessfully to withhold funds pending investigation of charges of modernistic teaching and permissive rules for students.[30] In Toronto, Baptist fundamentalist activist T. T. Shields conducted a long battle for control of McMaster University. Although fundamentalists south of the border rallied to Shields's aid, they could not succeed in ousting McMaster's relatively liberal professor of practical theology, the Reverend L. H. Marshall. In fact, Marshall was able to use Shields's heated rhetoric against him. When, after years of angry attacks, Shields compared himself to Christ, the "first fundamentalist," and called Marshall "Professor Pontius Pilate," the provincial Baptist convention kicked Shields out in 1926. Marshall retained his position.[31]

As did T. T. Shields in Toronto, many fundamentalists felt pushed out of power and influence in higher education by the later years of the 1920s. One group of students banded together to protect their status as both intellectuals and fundamentalists. The group, calling itself the League of Evangelical Students, formed in 1925 at Princeton Theological Seminary. Most of its early members were fundamentalist seminary students, although the group quickly branched out to include students from other types of colleges and universities. In the heated atmosphere of the later 1920s, especially as roiling educational controversies painted fundamentalism more and more explicitly as an anti-intellectual, reactionary movement, these students felt they needed to band together as isolated fundamentalists in the largely liberal and secular world of higher education. They also hoped to combat the growing consensus about the anti-intellectual nature of their movement. As would many other fundamentalists, this student group recognized their shrunken influence in higher education and hoped to project an image that combined steadfast fundamentalist theology with intellectual rigor. As their first constitution made clear, the student group believed that fundamentalism and modernism implied "mutually exclusive conceptions of the nature of the Christian religion."[32] Their group declared its activism to fight against "an agnostic or naturalistic Modernism."[33] Instead, the League devoted itself to promoting the "fundamental[s]" of the faith.[34]

In the opinion of League founders, other student groups such as the Young Men's Christian Association (YMCA) and the Student Volunteer Movement had succumbed to theological modernism.[35] They believed that campuses needed a group that would "stand alone" to promote the tenets of fundamentalist Protestantism in higher education.[36] As opposed to earlier fundamentalist college campaigners such as William Jennings Bryan, who had steadfastly hoped to regain cultural control of higher education, leaders of the League embraced their minority status on college campuses. As one member complained, "In many institutions evangelical Christianity is being completely ignored, in multitudes of others it is being definitely attacked."[37] Another noted that fundamentalists had been reduced to a "feeble folk" in higher education.[38]

This new status called, in the League's opinion, for an energetic campaign to project a different image for fundamentalists. One leader noted bitterly that "the enemies of evangelical Christianity claim to represent intellectualism and scholarship."[39] Another lamented that "those who call themselves the intelligent classes" had only "scoffing and satire" for fundamentalist belief and thought.[40] A third encouraged his fellow League members not to retreat in the face of accusations of intolerance. He recognized that fundamentalism had come to connote a "narrow-minded, bigoted, intolerant, or even unchristian" movement, but hoped that "constructive" evangelism on college campuses could overcome such shortsighted attacks.[41]

Nevertheless, the League of Evangelical Students never achieved the success its founders had hoped for. Perhaps because the image of fundamentalism on college campuses had been so bruised during the educational controversies of the 1920s, the pugnacious activism of the League did not attract the support it needed. Although by 1930 it claimed thirty-five chapters, including members at such prestigious schools as Harvard, the University of California at Berkeley, the University of Pennsylvania, and Oberlin College, and at such state institutions as the University of Washington and the State College of New York in Albany, the League folded by the end of the 1930s.[42] The death of leading supporter J. Gresham Machen in 1937 deflated much of the League's support. Even former backers such as Lewis Sperry Chafer at the Evangelical Theological College in Dallas had withdrawn their support by the end of the 1930s. The finances of the League dropped continuously throughout the 1930s until the group finally folded. In spite of this disappointment, the League managed to articulate a new position for fundamentalist scholarship that resonated with many 1920s-era fundamentalists. As we will see in Chapter 10, many fundamentalists beyond the ranks of college and seminary students felt a similar impulse to contest the image of fundamentalism as an anti-intellectual enterprise.[43]

As the fight for control of colleges and universities slipped away from fundamentalists, many fundamentalists sought to find or establish their own institutions friendly to the intellectual presuppositions of the movement. As public universities successfully defended their freedom to teach antifundamentalist doctrine, and fundamentalists accused more and more denominational schools of harboring modernist faculty or textbooks, many fundamentalist educators and activists agreed that fundamentalists needed some trustworthy guide to higher education. Where would students be safe from the teaching of evolutionary theory? Where would their fundamentalist beliefs be supported by classroom teaching and by strict parietal rules?

This need had been recognized as early as 1919, when Charles A. Blanchard of Wheaton College presented his survey of the state of Christian higher education to the founding meeting of the World's Christian Fundamentals Association (WCFA) in Philadelphia. Blanchard's committee on colleges recommended that the WCFA "make a list of such colleges, seminaries and academies as refuse to use text books or employ teachers that undermine the faith in the Bible . . . and in Jesus Christ as God."[44]

On the heels of this foundational meeting, Leander S. Keyser of Wittenberg College in Ohio and later of Evangelical Theological College in Dallas proposed a list of theologically safe textbooks. In the following years, Keyser published an annual list of acceptable books. Other fundamentalists contributed to the effort. James Gray, leader of the Moody Bible Institute, suggested his own list of reading material that fundamentalist college professors might use without fear. The drive to compile a list of fundamentalist-friendly colleges took longer to get off the ground. After years of conventions, surveys, and questionnaires, the Association of Conservative Colleges finally published a list of thirty-seven "safe" colleges in late 1927. Although there was still some disagreement over the finer theological points, all of these existing colleges declared their agreement with a nine-point fundamentalist creed. However, this short-lived group never became prestigious enough to attract many important players in fundamentalist higher education, including the leading Bible institutes and Wheaton College in Illinois. Its organization dissolved by the mid-1930s.[45]

This search for theologically and socially safe fundamentalist higher education in the years following the Scopes trial often spurred fundamentalists to found new schools of their own. In at least one case, however, fundamentalists sought to transform an existing denominational college into what William Bell Riley promised would be "the only strictly fundamentalist University in America."[46] In 1927, the Baptist Bible Union, a fundamentalist faction, offered to take over the struggling Des Moines University (DMU) in Iowa. The Reverend T. T. Shields, who had tangled in earlier years with McMaster University in Toronto, headed the university, assisted by Edith Rebman, the lay secretary of the Baptist Bible Union.[47]

As soon as he took control of DMU, Shields antagonized many faculty members. Many resented Shields's insistence on individual interviews with each faculty member. Shields hoped to vet the school of any modernism, evolutionism, or nonfundamentalist belief, and his interview style often veered into heated accusations. Every faculty member was also asked to sign a fundamentalist creed.[48] This process met with cheers from many national fundamentalist leaders. William Bell Riley delighted in the fact that professors could not "pussyfoot it" around the direct interrogation.[49] J. Frank Norris in Texas called DMU "A Modern Miracle," and the leaders of the Bible Institute of Los Angeles hoped that the new DMU was the start of a new trend in fundamentalist higher education.[50]

In spite of this support, the new Des Moines University soon found itself deeply troubled. For one thing, money was tight. Prominent fundamentalists had supplied plenty of editorials in support of the new effort in Iowa, but this moral support was not matched by monetary donations. Also, Shields and Rebman soon found themselves at odds with their newly hired president, Harry C. Wayman. Wayman resented Rebman's dictatorial attitude and worried about rumors of a sexual relationship between Shields and Rebman. Students were also unhappy. On one hand, many bristled at new rules implementing conservative social rules. For instance, many resented Shields's rule banning fraternities. On the other hand, some students complained that the atmosphere was not fundamentalist enough.[51]

It did not take long for open hostility to break out between Shields and Rebman on one side, and President Wayman on the other, supported by most of the faculty and students. In spite of their lack of popularity, Shields and Rebman controlled a majority of the board of trustees, and in May 1929, they crammed a resolution through a reluctant board, summarily firing all faculty members and requiring a personal reapplication through Rebman.[52] When news of this coup reached the frustrated students, they took matters into their own hands. As the board met inside the administration building, students gathered outside, throwing rotten eggs and chanting "Get Shields!" "Get Rebman!" "Break their necks!" Soon the students broke into the building, and most of the trustees took shelter in a storeroom. Some board members were not so lucky and found themselves dragged out of the building by students. The students emptied file cabinets, broke windows, and continued to shout for the necks of Rebman and Shields until the police—apparently dragging their feet in sympathy with the students' frustration—finally arrived.[53]

The police took the board members to jail for their own protection and Shields ordered the university closed. However, in many ways, Shields's abrasive personality had confirmed to many local residents the negative stereotypes from the Scopes trial. His aggressive self-promotion alienated many Des Moines residents already leery of the reputation of the fundamentalist movement as a whole. As a result, local police had sided with nonfundamentalist students. The district judge also ordered the university reopened for the rest of the school term to spite Shields and the aggressive fundamentalist movement he represented. Rebman soon left for China as a missionary, Shields returned to Toronto to his independent pastorate, and Wayman took a pastorate the next year in Kentucky.[54] After such high hopes, even the most optimistic of national fundamentalist activists did not come to Shields's defense. All of DMU's earlier supporters allowed the great experiment in explicitly fundamentalist higher education to fade as quickly as possible from the headlines.[55]

Other experiments in fundamentalist higher education had more success. Two of these stand out: Bryan College (originally Bryan University), founded in 1926 and opened in 1930; and Bob Jones College (now Bob Jones University), opened in 1927. In both cases, founders hoped to open institutions that would protect students from the mainstream cultural bias against their fundamentalist beliefs made so evident in the publicity surrounding the Scopes trial. The founders of Bryan College explicitly used the experience of Bryan at Dayton as their reason for existence. They wanted a college to honor Bryan's memory and perpetuate his beliefs. However, just as other fundamentalists struggled in the aftermath of the Scopes trial to define the new image of fundamentalism, so the founders of Bryan College wrestled with the issue of Bryan's life and legacy to the fundamentalist movement. Bob Jones Sr., founder of Bob Jones College, found himself and his mission energized in the aftermath of the trial. Unlike the founders of Bryan College, who struggled to come to terms with the new meanings of fundamentalism, Bob Jones Sr. proudly took up the challenge of new liberal stereotypes of the movement. His university would exist to defend and promote fundamentalism from unfair allegations of ignorance even as it implicitly accepted many aspects of the newly restricted definition of the movement.

William Jennings Bryan's dramatic death, on the heels of what fundamentalists viewed as his triumph at the Scopes trial, secured his place in the pantheon of fundamentalist martyrs. Fundamentalists quickly jostled to claim his legacy. One popular plan was the foundation of a college that would bear Bryan's name and continue his educational mission. Bryan himself would have probably suggested a secondary school, or academy, but his successors wanted a great and prestigious liberal arts college.[56] Most of them probably would have agreed with Bryan College's first president, George Guille, who wanted to see "a high-grade institution of learning" that would become "internationally known for its belief in the Bible as the inspired Word of God and for its devotion to the Lord Jesus Christ."[57]

Guille himself found that this lofty goal quickly attracted support from many prominent fundamentalists. It helped that Guille had many personal connections among the fundamentalist educational community. In 1925, he was nearing the end of a distinguished career as a Bible teacher at the Moody Bible Institute and as one of the cofounders of the Evangelical Theological College in Dallas. Guille used his contacts to muster an impressive list of prominent 1920s fundamentalists to support Bryan College: evangelist W. E. Biederwolf; J. C. Breckenridge from the Winona Lake Bible conference; J. B. Cranfill from Dallas; T. C. Horton from Biola; evangelist Bob Jones Sr.; D. S. Kennedy, editor of the *Presbyterian and Herald-Presbyter*; Austin Peay, governor of Tennessee; evangelists Paul Rader and Paul Rood; New York fundamentalist leader John Roach Straton; G. W. Taft, president of Northern Baptist Theological Seminary; and Florida real estate mogul and fundamentalist activist George F. Washburn. All of these men joined the national campaign committee in 1925 to raise money for a memorial college.[58]

Unfortunately for Guille and the other early leaders of Bryan College, lining up prominent fundamentalist supporters for the idea of a memorial college proved much easier than agreeing on the details of the new institution. For instance, fundamentalist activists disagreed about the proper location for the school. William Bell Riley strongly supported the idea that a "Great American University" would make the best memorial for Bryan.[59] But he argued forcefully that the proper location for such a university was Chicago, not Dayton, Tennessee.[60] Other supporters disagreed. George Washburn suggested a Miami location, but quickly sided with Bryan's widow Mary Bryan that Dayton would make a more appropriate location. In the end, Mary Bryan's prestige as the wife of the late leader won over Riley's opposition.[61]

Riley had protested that Dayton could be the home of an academy for younger students, combining a secondary education with some preliminary college-level academics. Chicago, in his opinion, had the advantage of proximity to Wheaton

College. The new university could build on this existing foundation, adding a Bryan Law School, a medical school, and other graduate departments. Riley agreed with Darien Austin Straw, a professor at Wheaton College, that the Wheaton location had convenient access to rail lines and roads. Plus, the Wheaton reputation had already been established. As Straw argued, "Our self-sacrificing opposition to popular evils has gotten us a confidence" among fundamentalists. Riley added that Bryan originally hailed from Illinois, making Chicago a natural choice. He even denounced the secret politicking of "certain Presbyterian men" who had tried to rally support for a Bryan memorial university in Dallas.[62]

In spite of these rationalizations, the vigor and vehemence with which Riley argued for a Chicago location suggests other motivations besides pragmatic considerations. One factor behind Riley's insistence on a Chicago university may have been simple ambition. Riley had long dreamed of a Chicago-based fundamentalist university and he argued that Bryan would have agreed if Bryan had only heard Riley's plan. The new university would presumably fall under Riley's influence. But Riley also likely felt a university in Dayton would become tied too closely with the stereotype of fundamentalism promoted at the Scopes trial. A Chicago university would be associated with a great northern metropolis, while one in Dayton might be linked in the public eye to the stereotype of the rough unlettered Tennessee dirt farmer.

Another stumbling block for the new Bryan College resulted from the touchy issue of doctrine. The new college proposed an eight-point doctrine, which all faculty and administrators were required to sign annually. This had become the norm for conservative Protestant schools. Leading evangelical schools, such as Gordon College, near Boston, had introduced mandatory faculty creeds in the early 1920s.[63] In Texas, Baylor College had instituted such a policy under unrelenting pressure from fundamentalist activists.[64] One of the first moves of T. T. Shields upon taking charge of Des Moines University was to install a similar creed.[65] The Moody Bible Institute of Chicago adopted such a faculty creed only in 1928. The Chicago leadership had been reluctant to discourage any potential faculty from accepting a position at the school. However, by 1928 it seemed prudent to join the movement toward mandatory creeds.[66] Charles Blanchard, president of Wheaton College, had long desired such a creed for his school, although the board of trustees did not find a need for one until after Blanchard's death in 1925.[67] Blanchard had argued such creed was necessary "because in every age, there are persons who profess to build upon the Word of God who proceed, having made such a profession, to deny all the essentials of Christian faith."[68] Like other fundamentalist educational activists, Blanchard agonized over the liberalization and secularization of many of the nation's most prestigious colleges and universities. Fundamentalists hoped that such ironclad creeds would secure their institutions from doctrinal change over time. They had seen how denominational institutions such as the University of Chicago had become headquarters of theological modernism and hoped that such change could be prevented.

In the case of Bryan College, then, the idea of a faculty creed was not a novelty, but the content of Bryan College's creed raised difficult questions about the proper definition of fundamentalism. The creed contained several standard fundamentalist tenets, including beliefs that the Bible was inerrant in its original "writings," in a triune God, in the sinful nature of humanity, and in the virgin birth and substitutionary sacrifice of Jesus. It also stipulated a belief in creation as described in the book of Genesis. However, it omitted any mention of the doctrine of the premillennial return of Jesus. This had long been a point of contention in the struggle to build a unified fundamentalist movement. The overwhelming majority of early fundamentalists believed in premillennialism; they held that Christ would return to usher in a thousand years of peace and earthly happiness. Postmillennialists generally believed that Jesus would return only after the end of that millennium. Some fundamentalists, such as William Bell Riley, insisted that premillennialism formed one essential component of true fundamentalism.[69] Other early fundamentalists carefully maintained the possibility that fundamentalism could include postmillennial belief. James Gray of the Moody Bible Institute (MBI) had argued for years that fundamentalism could include both. "The question of the second coming of Christ has been before the church in all the centuries," he argued in 1921, "and postmillennarians and premillennarians have been able to discuss it in love. . . . The real dividing line . . . is that of modernism versus evangelicalism. It is whether the Bible is a natural product, or a supernatural revelation from God."[70] Although Gray agreed that premillennialism was the position of most of the scholars at his Chicago institution, it was not a required tenet of the MBI's belief.[71] Other early fundamentalists agreed with Gray. Baptist editor Curtis Lee Laws argued that the fundamentalist movement must welcome all "premillennialists, postmillennialists, pro-millennialists, and no-millennialists," who were willing to stand "solidly together in the battle for the re-enthronement of the fundamentals of our holy faith."[72]

In practice, however, most fundamentalists insisted on premillennialism as a central fundamental of Christian faith. As their movement came under increasing attack, many fundamentalists hoped that premillennial faculty creeds could prevent any more confusion about the unchangeable elements of fundamentalism. The founders of Bryan College, however, found themselves in a dilemma. Rare among leading fundamentalists, William Jennings Bryan had espoused a traditional progressive postmillennialism. Including an insistence on premillennial belief in an institution meant to honor a man who did not believe in it would have been unthinkable. But rallying support and financial backing for a school that did not insist on a premillennial faculty creed proved difficult.

In fact, due to such disputes over doctrine and location, Bryan College had a very difficult beginning. Although the Bryan Memorial Association purchased land for the school in 1926, the first students did not meet until 1930, due to financial constraints. Those first students did not even meet in a Bryan College building, but ironically in rented rooms at the former Rhea County high school, where John Scopes had taught. The foundation for the main administration building was laid in 1927, but due to dire financial distress, that building was not completed until 1952. Although George Guille had easily managed to rally support for a memorial university in the months following Bryan's death, arguments about the nature of fundamentalism in the wake of the Scopes trial made it very difficult to raise the money necessary for such an expensive proposition. Fundamentalists who insisted on premillennial belief as a foundational tenet of fundamentalism hesitated to support such a school. Similarly, fundamentalists who hoped to distance themselves from the stereotypes promulgated about the Scopes trial worried that a fundamentalist college in Dayton would only serve to further cement that image of the fundamentalist movement.[73]

Bob Jones College was founded out of a similar sense that fundamentalists needed new higher educational institutions of their own. "Bob Jones College," Jones crowed, "is a protest against the atheistic, modernistic, Unitarian tendencies of much of our education."[74] However, unlike the fledgling Bryan College, founder Bob Jones Sr. embraced the fundamentalist label in spite of the power of the new stereotypes and built on that image to establish one of fundamentalism's most influential institutions. First opened in Panama City, Florida in 1927, it soon moved to Cleveland, Tennessee. Only in 1946 did the school become Bob Jones University, in its new home of Greenville, South Carolina.[75] Thanks to the publicity surrounding the issue of evolution and education, southern politicians scrambled to voice support for the well-known evangelist's new enterprise. Atlanta Congress member William Upshaw and Alabama state auditor S. H. Blan were among the dignitaries at the groundbreaking ceremony on December 1, 1926.[76] The new governor of Alabama, Bibb Graves, delivered the keynote address. In his speech, Graves paraphrased one of Jones's most popular sermons, thanking Jones for fighting the "curse" of "godless education."[77]

The new school succeeded beyond its founders' expectations. Part of the reason for its rapid success was its financially savvy fundraising appeal. Instead of soliciting donations directly, as the founders of Bryan College struggled to do, evangelist and school founder Bob Jones Sr. offered potential donors a chance to earn a profit on their contribution. The college began as part of a potentially lucrative land deal. Bob Jones Sr. lent his name to a real estate corporation led by New York–based developer Minor C. Keith. In return, Keith gave the new college a 470-acre tract, plus a promise to give the college 25 percent of profits from the sale of another three hundred acres, plus a free lot for a Jones family home. Bob Jones Sr. promised potential investors their money would "do four things: make a safe investment, line your pockets with profits, boost your community, and endow a college."[78]

Another advantage enjoyed by the new school was the deep pockets of founder Bob Jones Sr. His career as a fundamentalist itinerant preacher had earned him a handsome income. Throughout the 1920s, Bob Jones Sr. earned between $50,000 and $75,000 per year. This financial security allowed the evangelist to promise to pay $20,000 annually toward faculty salaries for the new college.[79]

Bob Jones Sr. was able to build on this financial nest egg by appealing to the experiences of fundamentalists throughout the decade. As we have seen, both professional and grassroots activists struggled unsuccessfully to assert control over public and denominational colleges and universities. Bob Jones Sr. repeatedly attacked existing institutions of higher education as leading fundamentalist students away from their faith. Jones used the term "shipwrecks" to describe the plight of these students educated out of their faith.[80] In his sermons and speeches in the 1920s, he often related anecdotes about such students. One unsettling story told of parents who scrimped and saved to send their daughter to "a certain college. At the end of nine months she came home with her faith shattered. She laughed at God and the old time religion. She broke the hearts of her father and mother. They wept over her. They prayed over her. It availed nothing. At last they chided her. She rushed upstairs, stood in front of a mirror, took a gun and blew out her brains."[81] Public colleges and universities, according to Jones, had become pits of damnation, fueled by "false educational philosophy." Such innovations as the elective system and the "self-expression, behavioristic idea of education" had destroyed the schools.[82] By the mid-1920s, Jones argued that the public schools were so destructive he "had just about as lief [sic] send a child to school in hell as to put him in one of those institutions."[83]

Unfortunately for fundamentalists, in Jones's opinion, most denominational schools were not much better. Jones attacked nonfundamentalist teachers at Christian schools in typically colorful language. "Bootleggers and harlots," Jones promised, "will stand a better chance in the day of judgment than the teachers in colleges called Christian which damn boys and girls from Christian homes."[84]

Jones offered fundamentalists an alternative. His new school, he promised, would pledge never to change its religious commitments. All of its faculty and board members would be required annually to sign a creed, to combat "all atheistic, agnostic, pagan and so-called scientific adulterations of the Bible." Plus, the creed could "never be amended, modified, altered, or changed."[85] Jones assured parents that their children would never lose their faith at the new college. He promised "fathers and mothers who place their sons and daughters in our institution can go to sleep at night with no haunting fear that some skeptical teacher will steal the faith of their precious children."[86] Further, parents could rest assured that the social lives of their children would also be strictly controlled. Women were not allowed off campus except with a family member or an "authorized chaperon." Students were prohibited from dancing, card playing, gambling, or listening to any music that might be considered "jazz."[87]

In addition to this appeal to the bitter experiences of fundamentalists with public and denominational higher education, Jones also built on the new image of fundamentalism as southern, reactionary, and, in a certain sense, anti-intellectual. He promised his audiences, "We are not going to deliver the South to this rationalistic, atheistic leadership."[88] In addition, he defended the image of fundamentalism as under attack from "worldly, highbrowed, snooty people."[89] Jones, like other fundamentalists in the wake of the Scopes trial, took umbrage at "an idea abroad among certain religious liberals that if a person believes in what is usually called the 'old-time religion,' he must, so to speak, have a greasy nose, dirty fingernails, baggy pants, and he mustn't shine his shoes or comb his hair."[90] Jones often attempted to refute the image of fundamentalism as opposed to all intellectual endeavor. Looking back, he remembered the early years of the school as an attempt to "build a college that would neutralize in the minds of the public the idea that culture does not go hand in hand with the old-time, conservative, Christian approach."[91] He promised his school would turn out more qualified professionals in all fields than comparable public or denominational schools. But Jones's understanding of intellectual and academic excellence differed from that of liberal or secular educators. For Jones, academic performance always came second to development of faith. Instead of training students to think critically about the world around them, as was the goal at many nonfundamentalist institutions, Bob Jones College promised to train students to strive to "think God's thoughts after Him."[92]

This appeal to established stereotypes about the fundamentalist movement worked. Unlike Bryan College, which struggled to raise money and attract students, Bob Jones College grew rapidly. In its first year, 1927, the college offered only a two-year associate's degree. By 1929, however, a full four-year program had been added. By 1932, elementary and secondary educational programs were included, as well as a master's degree program in religious education. The number of matriculated students also jumped quickly. From a first class of eighty-eight in 1927, enrollment quickly climbed to above two hundred by the next year. Enrollments continued to climb, even in the difficult years of the Great Depression.[93]

Bob Jones College and its founder Bob Jones Sr. were able to use new popular understandings of fundamentalism to attract support for a nondenominational fundamentalist college. Other schools had a more difficult experience in the later years of the 1920s. Bryan College struggled to frame a workable fundamentalist identity that straddled many of fundamentalism's boundaries. Des Moines University failed utterly to reform a denominational school into a fundamentalist powerhouse. And fundamentalist activists experienced mostly unsatisfying results in their campaigns to control local public and denominational schools. Most public colleges accepted some elements of fundamentalist pressure but continued to function as before. Denominational schools often rejected fundamentalist overtures to align more explicitly with fundamentalist Protestantism. In some cases, however, existing schools moved decisively into the fundamentalist camp, requiring faculty to sign new fundamentalist creeds and attracting students from proudly fundamentalist families. In every case, struggles over higher education in the aftermath of the Scopes trial used the new image of fundamentalism as tools to frame a new vision of higher education.

Further, the contests at colleges and universities had a singular impact on fundamentalist thinking. In spite of the stereotypes often promulgated about the fundamentalist movement, most fundamentalists in the 1920s cared deeply about the intellectual credibility of their movement. Colleges and universities embodied that kind of respectability. When major public universities rebuffed fundamentalist demands, many fundamentalists realized that they could no longer rely on the mainstream academic world. In response, many simply established their own institutions of higher education. As in the case of Bob Jones Sr., many 1920s-era fundamentalists insisted that their new schools would be "to the educational world what a demonstration farm is to the agricultural world. We are proving in our institution that it is possible to be thorough in scholastic work and have a happy, contented student body in this modern age, and still hold to the old orthodox religious position of our fathers." If fundamentalists could no longer assert control over the teaching at mainstream colleges and universities, at least they could establish schools in which their cultural demands could become the cornerstone of a new, independent, and explicitly fundamentalist system of higher education.[94]

CHAPTER 8

Fundamentalists, Bibles, and Schooling in the 1920s

In June of 1925, as Tennesseeans braced for the showdown over the teaching of evolution in Dayton's upcoming Scopes trial, the state board of education quietly strengthened the state's existing law about mandatory Bible reading in public school classrooms.[1] Since 1915, Tennessee law required public school children to read from the Bible every day in class.[2] In addition to requiring public school students to hear passages from the King James Bible, the new law offered school credit for Bible study. Not only would every student hear the Bible, but some students could earn school credit in Tennessee schools for studying the Word of God.[3]

Unlike the Butler Bill, which had banned the teaching of evolution in Tennessee schools, this expanded Bible law attracted little national notice or controversy. One reason for this lack of attention was because the Bible law was not seen as a purely fundamentalist measure, nor was it widely associated with the antievolution movement. The law put control of the Bible class curricula in the hands of a board composed of Protestants, Catholics, and Jews.[4] Although most fundamentalists ardently supported such laws, the campaign to increase public school students' exposure to the Bible claimed much broader support than so-called antievolution measures. Unlike the whirlwind of media attention on the Scopes trial and the antievolution movement, the drive to mandate Bible reading in public schools remained much less controversial throughout the 1920s. It did not force conservative Protestants to embrace or disavow the controversial image of fundamentalism. Conservatives who did not want to associate with fundamentalism given its new post-Scopes trial connotations could still enthusiastically support mandatory-Bible campaigns.

Indeed, many liberal Protestants also participated in the drive to require Bibles in public schools. Even when liberal Protestants opposed such laws, however, they usually did not fight against them as vigorously as they did against antievolution laws. Fundamentalists of all kinds worked energetically alongside nonfundamentalist allies to make sure every public school student read or heard the Bible during his or her school day. In marked contrast to their so-called antievolution activism, these fundamentalists were not forced to accept or reject constricting public stereotypes about fundamentalism in order to do so. Because the Bible campaign did not have to struggle with the changing definitions and meanings of the fundamentalist movement, it continued to motivate a wide coalition of Protestant activists throughout the 1920s.

Many of those activists were motivated by their perceptions of change in American education and culture. Changing patterns of social morality prompted many Protestants to act on their desire to keep or return Bibles to America's public schools. Many conservative and fundamentalist Protestants argued that the only way to heal American culture was to expose American schoolchildren to the salutary effects of the King James Bible. This insistence hoped to give previously implicit beliefs about the moral nature of schooling the explicit force of law. As with the "antievolution" campaign, in which evolution came to represent a brace of doctrines offensive to fundamentalists, the Bible came to symbolize an educational panacea in the minds of many conservative activists. For fundamentalists, Bibles generally represented a supernatural tool to save souls. Other conservatives often put more emphasis on different positive results of Bible reading in schools. Exposure to the Bible, many nonfundamentalist conservatives believed, would instill patriotism and an urgently needed sense of morality in America's schoolchildren. Fundamentalists and other conservatives agreed on the healing potential of Bibles in schools, even if they usually emphasized different reasons for that potential.

This wide support, however, did not mean that the controversy over the role of the Bible in public schools had disappeared. Fundamentalist and nonfundamentalist Bible supporters alike remembered the bitter history of struggle between Catholics and Protestants over the issue. Since the mid-nineteenth century, Catholics had protested against the reading of the King James Version of the Bible in school exercises, since it was often accompanied by anti-Catholic commentary.[5] For most of American history, however, Protestant educators of all stripes agreed that public schools were appropriate places to expose children to the Bible. Although Catholics and other minority groups continued to object, Protestants generally viewed the Bible as appropriate for public schools. The reading of the Bible, many felt, did not constitute sectarian instruction. Rather, mainstream Protestants believed the reading of the Bible and the recitation of the Lord's Prayer belonged to all Christian denominations.[6] Even Horace Mann, an indefatigable campaigner against sectarian education in public schools, pointed out in 1847 that his Massachusetts Board of Education had worked to "promote . . . encourage, and . . . to *direct* the daily use of the Bible in school." Mann, like his contemporaries, considered the reading of the Bible without comment to be an entirely appropriate part of public schooling.[7]

In spite of continuing battles, the practice of daily Bible reading in public schools had remained the norm throughout most of the nineteenth and twentieth centuries. Only twelve states did not allow reading from the Bible in public schools during the 1920s.[8] Nationwide, the legal status of school Bible-reading only changed in 1963, when the Supreme Court ruled that such religious practice violated the Constitution.[9] During the 1920s, the debate usually did not center on whether or not the Bible could be read in a public school but instead on the method of that reading. Some Protestant educators and activists attempted to satisfy non-Protestants by suggesting that the Bible could be read daily without comment and that students with religious objections could simply be exempted. Even some fundamentalists who were eager to promote Bible-reading in public schools agreed that such reading could not include instruction in doctrine. Many of them agreed with Clarence Benson, director of religious education at the Moody Bible Institute, who glumly accepted "the futility of compelling our public schools to teach the Bible."[10] William Jennings Bryan, who became a staunch supporter of Bible laws for schools, originally accepted the premise that "the defense of the Bible is not permitted in schools supported by taxation." Many fundamentalists in the early part of the 1920s followed this line of reasoning. They agreed that public schools should not teach biblical doctrine, but they insisted that the Bible could still be read neutrally without inflaming sectarian differences. Others pushed for a more traditional Protestant teaching of doctrine, along with a daily reading of Scripture.[11]

During the 1920s, fundamentalist educational activists disagreed among themselves about which of these goals was more practical, but most favored some form of Bible reading in public schools. Throughout the decade, these activists wanted every public school student to hear the Word of God at some time during the school day. They helped pass state laws requiring daily Bible readings, or as in Tennessee, allowing students to earn school credit with Bible study. Beginning in 1913, several states passed mandatory-reading laws, including Pennsylvania (1913), Tennessee (1915), New Jersey (1916), Alabama (1919), Georgia (1921), Delaware (1923), Maine (1923), Kentucky (1924), Florida (1925), Idaho (1925), and Arkansas (1930).[12]

Just as other school laws had, these new state laws represented only the most successful examples of a much broader movement. Many other state legislators tried and failed to pass similar bills. For example, the successful Bible-reading measure in Florida came two years after state lawmakers squelched a 1923 bill.[13] In Ohio, a bill passed both houses of the legislature in 1925 only to meet the veto of Governor Vic Donahey.[14] In Washington in 1926, a proposed constitutional amendment to permit school Bible-reading passed the House by a wide margin, only to be defeated in the Senate.[15]

As with other fundamentalist state legislation, successful Bible laws often relied on zealous state legislators who refused to accept defeat. West Virginia lawmakers, for instance, had twice defeated mandatory Bible laws in 1923.[16] In 1927, Representative W. A. Street of Barbour County began his legislative session by introducing a mandatory school Bible bill. Street's bill eventually passed the House, only to be defeated in the Senate.[17] The savvy lawmaker, however, did not rely solely on the passage of a single bill. Later in the session, Street also introduced a popular bill to provide free textbooks for the state's elementary schools. Only after the bill had received a positive recommendation from the Committee on Education did Street attempt to amend the bill to include both Bibles and traditional McGuffey's Readers into the list of approved textbooks.[18] Just as Street had introduced multiple bills to ban evolution and other antifundamentalist ideas from West Virginia's public schools, he worked diligently to get Bibles into those schools by any methods necessary.[19]

Even when statewide laws failed passage, fundamentalists and their allies pressured local school boards and administrators to adopt Bible curricula. As this chapter will describe, fundamentalists also continued a tradition of aggressive Bible-based missionary work. Much of this outreach in the 1920s targeted public schools. For instance, throughout the decade, fundamentalists distributed hundreds of thousands of Bibles and fundamentalist tracts directly to public school pupils and teachers. Some fundamentalist activists sponsored programs that taught students a Protestant fundamentalist interpretation of the Bible outside of public schools. For some fundamentalists, this entailed volunteering to direct "release time" programs. Elsewhere, fundamentalist activists supplemented children's education with private clubs and classes after school or on weekends.

Many of these campaigns enjoyed great success. Part of that success derived from the fact that these Bible campaigns never became embroiled in the struggles over the meanings of fundamentalism that had resulted from the so-called antievolution campaign. Although fundamentalists led many of these disparate Bible campaigns, the campaigns never became associated in the popular mind as a distinctly fundamentalist program. Nevertheless, some critics of the fundamentalist movement tried to associate the Bible campaign with the new stereotyped image of fundamentalism. Harbor Allen, the publicity director for the American Civil Liberties Union, reported with alarm, and without accuracy, that fundamentalists such as those in the Supreme Kingdom had successfully passed mandatory-reading laws in several states in 1926. Allen hoped to draw public attention to what he saw as yet another fundamentalist assault on civil liberties and liberal values. However, the Bible drive never achieved the same notoriety and identification with fundamentalism as the "antievolution" campaign had. Conservative and even some liberal Protestants felt comfortable promoting Bible-reading in public schools, without facing accusations of ignorance and intolerance.[20]

Such nonfundamentalist Protestants had a long tradition of aggressively promoting Bible-reading in public schools. National groups included the American Bible Society, the American Tract Society, the National Reform Association, the American Home Missionary Association, and denominational missionary boards. Some new groups also joined the struggle during the 1920s, most notably the militantly Protestant Ku Klux Klan. Although all Protestants agreed in principle that Bibles could save the souls of children, most of these nonfundamentalist groups emphasized arguments about the social benefits of Bibles. They argued that American society needed to keep Bibles in public schools for practical reasons rather than spiritual or theological ones. These mainstream Protestant organizations generally did not stress the soul-saving aspects of mandatory Bible reading. Rather, they pointed out that without a solid grounding in biblical morality, society would descend into chaos. The Bible, they argued, provided a necessary dose of social hygiene, cleansing society's tendency toward sin and decline.

Although some of the nonfundamentalist groups that pushed for passage of Bible laws during the 1920s were local ad hoc coalitions, many had roots stretching back to the Jacksonian "benevolent empire." Older groups such as the Young Men's Christian Association (YMCA) had moved away from aggressive Bible distribution by the end of the nineteenth century. The YMCA emphasized physical education in part because they felt they could not compete with new public libraries.[21] Other "benevolent empire" groups such as the American Bible Society (ABS) continued to distribute Bibles. In 1926, for instance, the ABS claimed to have distributed almost ten million Bibles.[22] However, even this impressive output reflected a loss of cultural clout for the organization since its heyday in the mid-nineteenth century.[23]

Both groups, like other nonfundamentalist Bible campaigners, argued that daily Bible reading would strengthen the moral backbone of public school students. This argument had a venerable history. Since the early nineteenth century, social reformers had campaigned to increase reading of good literature and books generally, believing it could ensure social harmony. George Ticknor, for instance, one of the founders of the Boston Public Library, advocated a free library on the grounds that rebellion and chaos would result from a public deprived of books.[24] By the 1920s, many nonfundamentalists used similar arguments to push for Bibles in public schools. Luther A. Wiegle, professor of religious education at Yale, bemoaned the lack of religion in the public schools, which would lead to crime and create an entire generation without any moral compass. The failure of society to indoctrinate its young in biblical morality, Wiegle argued, "imperils the future of the nation itself."[25]

The Ku Klux Klan aggressively promoted Bible reading in schools during the 1920s. The relationship between the Klan and fundamentalism was a complicated one, as described in Chapter 2. Although the new Klan had won sympathy from a few prominent fundamentalist leaders, most fundamentalists kept their distance. Moreover, the Klan welcomed all kinds of Protestants into its ranks, including some theologically liberal Protestants who were often the most ardent foes of the fundamentalist movement.[26] The Klan tended to emphasize different reasons for installing Bibles in public schools than fundamentalists did. Many Klan leaders advocated mandatory reading of the Bible as an anti-Catholic, anti-immigrant panacea. R. H. Sawyer, a national itinerant lecturer for the Klan, advocated the return of the King James Bible to public school classrooms "as it was in the old days." This simple school reform, Sawyer argued, would be enough to return political control to native-born white Protestant men. "Within the next few years," he promised, this measure could ensure that "only native born Americans rule the government instead of foreigners."[27] One local klavern leader—an "Exalted Cyclops," in Klan terminology—told the 1923 conference of Klan leaders that the Catholic Church was seeking to destroy the nation's public schools. One way to counteract this influence, he argued, was to make sure every public school student learned the Bible: "The Bible must be read and explained to [public school students] daily during their early school years. The Knights of the Ku Klux Klan believe that the free public schools should be the vehicle for this Bible reading and instruction and that no atheist, infidel, skeptic, or non-believer should be allowed to teach in the public schools. The Klan does not contend for sectarian instruction in the Bible, but asks that it be read and explained from the broad viewpoint of its divine origin and inspiration."[28] John Galen Locke, leader of the Colorado Klan, emphasized that one of the main goals of the organization was to "place the Open Bible in the Public Schools of America."[29] Sometimes local Klan groups exerted political pressure on local officials. Many women's Klans in Indiana, for instance, pressured their local school boards to adopt compulsory Bible reading in public schools.[30]

Other local Klans sought to use a carrot instead of a stick. In Anaheim, California, a contingent of robed and hooded Klansmen presented a local pastor with a cash donation and a letter of gratitude, thanking him for placing Bibles in every public school classroom in the city.[31] In Akron, the local klavern donated Bibles and flags to the local public schools, and the school reciprocated by offering an elective course in Bible study.[32] This tactic was fairly common. According to historian Kathleen Blee, almost every local women's Klan in Indiana attempted to donate Bibles and flags to their local public schools. They often donated multiple copies as well as "Stories from the Bible" and placards bearing the Ten Commandments. Many of these Indiana activists met with resistance. While the women's Klan of Coal City was allowed to conduct an elaborate ceremony during the intermission of a school play, during which they donated a set of Bibles to the school principal, the Terre Haute women's Klan was denied permission. The Terre Haute school board decided that the ceremony, complete with robed Klanswomen, was not appropriate and the Klanswomen refused to make the donation in civilian attire.[33]

In many other cases, local Klan organizations put political pressure on elected officials to pass mandatory-reading laws. In Ohio, for example, the 1925 Buchanan Bible Bill to mandate Bible reading in public school was widely viewed as a test of the "political power of the Ku Klux Klan." After Governor Donahey vetoed the bill, the Ohio Klan vowed that this meant "political oblivion" for the Governor.[34]

Many grassroots activists likely perceived no conflict between fundamentalist and nonfundamentalist Bible campaigns. What differences there were often tended to appear as a difference in tone and emphasis. For instance, one article reprinted in the national Klan newspaper in 1923 insisted that schools should be "run by the free and untrammeled powers of Protestants," a goal few fundamentalists would dispute, at least privately. But the ultimate end of that Protestant education, for the Klan, was patriotic instead of theological. The Bible, for many Klan activists, was a necessary book to help impart patriotic feeling, to raise a new generation of "true American citizen[s]." For fundamentalists, Bibles in schools ought to be there for their own sake; the goal was not citizenship, but saved souls.[35]

A second nonfundamentalist organization that focused on passing mandatory-reading laws was the National Reform Association (NRA). This group, though composed largely of Protestant clergymen, differed from the fundamentalist movement in that it opened its doors to both liberals and conservatives, although most of its support came from conservatives. Further, the group did not stress the fundamentalists' belief that the Bible was an inerrant book, or that it had supernatural power to save souls. Its argument for Bibles in public schools was one both liberals and conservatives could agree on: the Bible in schools solved a social dilemma. W. S. Fleming, a former Chicago pastor and member of the national "field force" for the NRA, claimed that Bible laws for public schools would enable society to maintain basic morality.[36] Fleming pointed out that most states gave Bibles to prison inmates. Why not skip the middleman, he asked, and deliver the Bibles to the schools? If Ohio had followed this suggestion in 1925, he recalled, "as her neighbor, Pennsylvania, did, with the same result, more than half of her present 9,310 convicts would now be law-abiding citizens."[37]

In the opinion of the NRA, release-time programs for Bible study were not enough. Most of the benefits from released time, the NRA believed, went to those students who needed it least: affluent urban kids who already attended church and Sunday school with their families. Release-time programs could only function in districts wealthy enough to attract volunteer denominational teachers. Schools in poor urban and isolated rural districts could not get such volunteer help.[38] Instead, the NRA supported programs such as the one initiated by the school board of Chattanooga, Tennessee, in 1922. In this program, the Bible was taught as an elective course in school buildings on school time, from fourth grade through middle school. In high school, the course was taught daily. In the Chattanooga system, a "Bible Study Committee" chose the Bible teachers. Each of these committees was composed of members from the YMCA, the Young Women's Christian Association (YWCA), a local Pastors' Association, the Parent Teacher Association, and the school committee. According to the NRA, this was a popular program, enrolling "nearly one hundred percent" of lower school students, and avoiding local controversy by including all Protestant denominations on the committee.[39]

The NRA scored a lasting success with its role in the Arkansas debate over a statewide mandatory-reading law in 1930. In this case, a local ad hoc coalition, calling itself the Moral Culture League of Arkansas, put the Bible Reading Act on the 1930 ballot though an initiative drive. The initiative passed readily, with support from most Protestant clergy, both liberal and conservative, and opposition mainly from urban Catholics and Jews. At least one liberal Presbyterian, the Rev. Hay Watson Smith, publicly opposed the measure, but among Protestants he formed a lonely minority of one. As the campaign continued, the NRA sent Fleming to speak in Little Rock a month before the election. Fleming's concerns with rising crime rates and the need for biblical morality echoed the social-hygiene themes in the pamphlets already disseminated by the local Moral Culture League. Unlike the antievolution law, which had only passed in 1928 after an angry and divisive statewide debate, this Arkansas law passed with little controversy.[40]

Other local campaigns for mandatory-reading laws encountered more resistance. In Texas, state legislators repeatedly introduced but were unable to pass a mandatory-reading bill. Even though the measure to begin each school day with a reading from the Bible, without teacher comment, was defeated in 1923 and 1925, a Texas school survey in 1927 found that many Texas schools read from the Bible. Of 547 schools across the state, 259 reported formal Bible reading, even though 370 felt that the legislature should not require it.[41]

Addressing school issues locally was a common response to deadlock or inattention to the issue at the state level. Schools, after all, were locally controlled and mostly funded through local property taxes. In North Carolina a split developed in the coalition to support Bible laws between Presbyterians and Baptists. In this case, Presbyterians supported the law, while Baptists opposed it, largely due to their strong tradition of opposition to state interference in religious matters. The legislature was unable to agree on any statewide law. In spite of this, by 1925 five North Carolina cities had authorized the establishment of Bible courses in their public schools.[42] Across the nation, many other small cities and towns followed suit. Middletown, Ohio (population 32,000) and Burlington, Iowa (26,000), passed municipal Bible-reading laws in 1924 and 1922, respectively.[43]

Because of the widespread popularity of mandatory Bible-reading measures, campaigns to pass such laws often attracted much less attention from the general public during the 1920s than the campaigns to ban evolution. Fundamentalists, too, differed in their levels of interest in mandatory-reading laws. Many prominent fundamentalists took little notice of the Bible laws. Others, however, made it a central point of their educational activism. To further complicate the picture, even those fundamentalists who made the issue central to their educational campaigns had differing reasons for doing so. All fundamentalists may have agreed with one Presbyterian commentator who argued that America "must restore the Bible to its historic place in the family, the day school, the college and university, the church and Sabbath-school, and thus through daily life and thought revive and buildup her moral life and faith, or else she might collapse and fail the world in this crucial age."[44] However, fundamentalists also often emphasized different reasons for mandatory-Bible laws. They agreed with more liberal Protestants that such laws would promote social stability. More prominent in fundamentalist rhetoric, however, was the impact Bibles would have on the souls of school children. Reading the Bible, fundamentalists argued, would save students from all the spiritual dangers of the modern age, including atheism, secularism, and even liberal Protestantism. Although they were willing to work with more liberal allies to promote Bible-reading in schools, fundamentalists assumed the practice would bring students to a fundamentalist interpretation of that Bible. For many fundamentalist activists, then, it was critical that the Bible be read in the correct manner. As M. H. Duncan, author of *Modern Education at the Cross-roads* (1925), argued, just having Bibles in public schools might not be enough. Duncan believed that the teacher must embrace "the Bible as the very Word of God." Otherwise, the full impact of the message would be lost.[45]

Bryan's approach to the issue of Bibles in public schools was very similar to his crusade to ban the teaching of evolution and atheism. Just as with evolution, the seasoned politician showed an eagerness to cooperate with nonfundamentalist allies. Bryan assured prominent Catholic, Jewish, and liberal Protestant colleagues that his advocacy of Bibles in the classroom would not lead to fundamentalist tyranny. To further his goal, Bryan used his influence with state lawmakers to help pass the 1925 Florida law to "make the reading of the Bible compulsory in public schools."[46] The danger, Bryan felt, lay with the current trend toward "perverted education." America's schools were moving students away from morality, in marked contrast to their original mission, and in contrast to the will of America's parents. Since he believed that "intellects are dangerous to society unless properly directed," Bryan advocated mandatory Bible reading, with exceptions made only if parents protested. He had promised his Catholic and Jewish friends that each group would receive equal treatment in the public schools. Bryan liked the release-time model, since it would guarantee minority rights yet recognize the necessity of religious instruction in public schools.[47]

Other fundamentalists did not have the same political clout as Bryan, but they worked just as hard to promote mandatory-reading laws. The Bible Institute of Los Angeles (Biola) promoted the idea that the Bible was a necessary component of public education. Unlike Bryan, the leaders of Biola worked in a state that had legislated Bibles out of public school classrooms. In addition to many direct-action programs to install the Bible in those classrooms, the leaders of Biola also strenuously advocated a legislative reconsideration of the California state law that had declared the Bible a sectarian book. In 1920, T. C. Horton, then editor-in-chief of Biola's *King's Business*, outlined the school's position. As part of their attempt to make "the Country Safe for the Children," Horton argued that he did "not want the Bible *taught* in the schools, but we do want the Bible honored and read."[48]

The Biola community gave many reasons to support a new look at the state law. They argued, in part, that democracy demanded it. In the words of one Biola writer, California must not let "a comparatively small branch of the nominal Christian church and a few Jews and foreigners" force the Bible out of the public schools. If the majority of Californians supported the mandatory reading of the Bible in schools, then it was every citizen's patriotic duty to pass new laws.[49] Furthermore, in the opinion of many leading Biola voices, education itself was impossible without the Bible. An understanding of literature, philosophy, history, and even physical science all required a thorough knowledge of the Word of God. In the words of John M. MacInnis, who became dean of Biola in 1927, it was "absolutely impossible for any teacher to scientifically teach American history in the public schools and keep the Bible out of the schools."[50] Even if a teacher could teach without the Bible, it would result in morally stunted youth, according to Biola writers. Many appealed to history. They cited the cases of the Soviet Union and the secularized schools of Bismarck's Germany as proof that Godless schools produced amoral, militaristic, criminal youth. If morality were left in the hands of secular teachers, it would "never rise above the law of the school room."[51] In other words, the only compelling moral education was one based on the Bible. To the Biola community, the choice was clear: either accept "this system of Godless, Bible-less education" or fight for a change in California law. In spite of this strenuous advocacy, the Biola community did not have the political clout to bring about legal change. Instead, as we will see, most of their success came with their programs to circumvent the law by delivering Bibles and Bible education directly to public school children.[52]

Isolating the relative influence of fundamentalists in passing mandatory-reading laws during the 1920s is complicated by the local nature of reform efforts, which were promoted in different communities, states, and regions. In most cases, fundamentalists formed one segment of local and statewide coalitions that pushed for such laws. Most crucially, unlike the antievolution fight, the Bible issue never became equated with the fundamentalist movement. Supporters of Bible laws did not have to decide whether or not they were fundamentalists. Liberal Protestants, conservative but nonfundamentalist Protestants, and fundamentalist Protestants did not split over the issue, the way they had over the issue of evolution. Although there are some exceptions, such as Bryan's successful lobby for a Florida law, most fundamentalists did not play a leading role in organizing legislative action in favor of Bible-reading. Much more commonly, fundamentalists took direct action to deliver Bibles into public schools, regardless of state law. Some efforts were very modest; some never got past the planning stages; but other activists worked to ensure that every public school student heard the Word of God at school.

Many local fundamentalist groups sought to deliver Bibles directly to students, and thus avoided legislative fights or much publicity. Kansas Presbyterians, for example, exemplified this quiet approach by compiling a "teachers' manual of Bible lessons." This manual was delivered free of charge to any rural schoolteacher willing to give students at least one hour of Bible instruction.[53] Another small fundamentalist group, based in Princeton, New Jersey, promised public school students a free New Testament if they read all four gospels and promised to pass them on to another student and encourage him to read it. By 1927, this group claimed to be operating in over two hundred East Coast cities, a claim that is difficult to document.[54] Some schemes to deliver Bibles directly to public school students, however, never got off the ground. A group of Philadelphia-based evangelists calling their program the "Million Testaments Campaign for Students" announced a grandiose "New Campaign" to deliver a million Bibles to students across the United States and Canada. One fundraising advertisement warned, "Like a blast from the pit, infidelity and atheism are sweeping through out colleges and schools." In spite of such dire warnings, this ambitious campaign never amounted to anything beyond attractive advertisements.[55]

Some direct-action Bible campaigns, however, were enormously successful. Bible educator Elizabeth Evans began a campaign in the late 1920s to deliver Bible education to public school students in New England. Unlike nonfundamentalist Bible education groups such as the Ku Klux Klan and the National Reform Association, Evans' drive began explicitly as part of the fundamentalist movement. Although by the 1980s she had come to prefer the term "Evangelical" to describe her work, her organization, James Elwin Wright's New England Fellowship, clearly identified with the fundamentalists during the 1920s. At an early meeting, for instance, Wright featured national fundamentalist leader William Bell Riley prominently on his flyer. Wright knew that this advertisement would prompt any Protestant in the 1920s, "if they know anything about Christian circles at all," to associate his new group with the fundamentalist movement.[56]

What distinguished Evans' work from nonfundamentalist Bible promoters in the 1920s was her desire to use Bibles in schools to convert students to a specifically fundamentalist theology. New England cities were served by hostile "modernist pastor[s]," so she went instead to isolated rural schools that lacked any access to church life or Sunday schools. Although at least one state superintendent attempted to pressure her to teach a more ecumenical curriculum, Evans refused. She agreed to travel to isolated schools but only if she were allowed to teach "the whole . . . Bible and the life of Christ." By the 1940s, Evans' direct-action Bible campaign expanded to over twenty thousand children. By then, she was producing her own Bible curriculum, training volunteer teachers, and even producing recruitment films to attract new "missionaries."[57]

The fundamentalist community of the Bible Institute of Los Angeles conducted similar campaigns during the 1920s to deliver the Bible into public schools in Southern California. Although the "B.I."—as it was known among the community during the 1920s—was unable to change California's state law against Bible teaching in public schools, they circumvented the law by opening several different types of clubs that reached out to public school students. The Nuntius clubs, for instance, sought to attract middle- and high-school age boys from around the Los Angeles area. With a program that combined Bible verse memorization with athletic competition, these clubs hoped to teach local boys to "fight temptation, keep the body and mind clean, and to save them for Christ and the church." By 1928, the clubs attracted hundreds of boys to local track meets and picnics.[58]

The B.I. had even more success with similar clubs for girls, the Euodia clubs. Named after the hardworking early Christian woman discussed by Paul in the book of Philippians, these Euodia clubs were organized by school, so that each public school had its own club. The clubs met in private homes or in fundamentalist churches close to the public schools. The Euodia clubs did not stress physical education as the Nuntius clubs did. Instead, they taught a fundamentalist interpretation of the Bible, hoping to teach girls to remain "fragrant for Christ," and worked at "fortifying them against the temptations which confront the youth of today." By the end of the decade, almost one thousand girls participated in these clubs.[59]

The Bible Institute of Los Angeles also worked to teach the Bible to public school students during the 1920s though a "Children's Garden" program. Sophie Shaw Meader initiated this program in 1924 to reach out to students nationwide. The idea was simple. Every three months, fundamentalist students at public schools would receive, in the *King's Business*, graded Bible study questions. Students who completed these assignments and sent them to Meader were awarded prizes. Several public school teachers used the materials in their classrooms, including respondents from Kansas, Idaho, and Michigan, as well as California. Other respondents used the program as part of a religious home-schooling curriculum. This correspondence course only lasted a few months, after which Meader transformed the "Children's Garden" into a pull-out section of the *King's Business* magazine, meant to be interesting to children.[60] The B.I. also trained many public school teachers. Many alumni of the B.I. went on to missionary work around the nation and around the world, and some alumni sought to spread the fundamentalist message by teaching the Bible in public schools. In some states, that was not controversial. Lura Faye Hixson (Class of 1925), for example, moved to Graham, Texas, where she took a job teaching the Bible in local public schools. Since many of the B.I. alumni settled in California, their missionary work in public schools forced them to subvert the law actively. Elizabeth Hunter (Class of 1925) worked in the "crowded, foreign districts" of Los Angeles. A public school teacher, Hunter remained true to the B.I. ideal and saw her main role as using the Bible in school to promote conversion to evangelical Protestantism.[61] Other alumni public school teachers were more careful to respect the letter of the California law. Ruth A. McClain, a 1918 B.I. graduate, also taught in the Los Angeles public schools. Unlike Hunter, she refrained from openly teaching the Bible during the school day, although she continued to "pray that the time may soon come when the Bible shall not be barred from the public schools." In order to promote the Bible among her students, she taught a Bible class for her students in a private home near the school, after school hours.[62]

By far the largest direct-action campaign to place Bibles in public schools was conducted by the Moody Bible Institute of Chicago (MBI). William Norton, director of the MBI's Bible Institute Colportage Association (BICA), began the school Bible drive in 1921, after a missionary trip to the southern Appalachian region. As far as Norton could tell, many of the "mountaineers" were nominally Christian, but rarely saw a Bible or read one. "To reach these people quickly," Norton argued, "I am convinced that it can be done most efficiently . . . through the public schools. . . . A great majority of the teachers are ready to cooperate."[63]

Norton soon established a series of "book funds" to help pay for the massive public school distribution campaign. To raise money, he stressed two contending themes: opportunity and threat. BICA fundraising brochures described the unique opportunity presented by poor but eager public school children, so desperate for reading material that they avidly read the Gospels and Colportage library books. The flip side of this opportunity, however, was that these vulnerable and open-minded children could be attracted to some of the "unwholesome and evil books" that had begun to flood the mass market.[64]

The 1924 annual report of the Moody Missionary Book Funds bemoaned the fate of the "Mountain dwellers," and highlighted the spiritual work to be done: "Back in the more inaccessible regions of the Southern Mountains and in the Ozarks, *another generation of bright young folks are growing up with little or no Christian influence.* These boys and girls are eager for such mind- and soul-awakening literature as the 'Moody books.' . . . *Surely you will agree that such a remarkable opportunity to transform young lives for time and eternity should not slip by on account of a lack of dollars.*"[65] Elsewhere, members of the Moody Bible Institute community were informed that the Southern Appalachian public schools were "*THE MOST WONDERFUL SOUL-WINNING OPPORTUNITY IN AMERICA! JUST THINK OF IT!* The Word of God impressed upon these young minds everyday in the week! Then, too, double the number of home folks are also reached with the Gospel."[66] These depictions of the uniquely needy "mountaineers" ran through both the fundraising literature and the internal reports of the BICA. Just as Elizabeth Evans focused her Bible education on isolated rural schoolchildren so that she could freely deliver her fundamentalist message, the Book Fund constantly emphasized the Appalachian schoolchildren's poverty as the best insurance that they would appreciate the free literature and benefit from its evangelical message.

A variant of this theme was the willingness of teachers in the "Mountain" regions to use Colportage books and Scripture memory programs in their classrooms. One fundraising brochure trumpeted, "Thousands of mountain and pioneer school teachers are ready to use Scripture portions and Moody books as a means of bringing their pupils to Christ. *What an opportunity!*"[67] Another brochure lamented the fact that teachers were even willing to "drill their pupils on the great salvation truths of the Bible, but they lack even the Gospel of John with which to begin."[68]

However, this opportunity for evangelism was in continual tension with the growing threat of modern secularism. The Missionary Book Fund of the BICA continually stressed the idea that the "mountaineers" were no longer so isolated that they were entirely safe from the spiritual threats of modern life. In 1896, just two years after the BICA's founding, one informational brochure explained that the BICA had been "FOUNDED . . . to help stem the flood of vicious literature that is now in circulation."[69] In 1921, one of the five official goals of the BICA was "to counteract the tide of unwholesome and evil books and papers which is flooding the country."[70]

The method was relatively simple and followed a tried-and-true Sunday-school model. Every three years, teachers were contacted and offered a set of books. If they agreed, the Missionary Book Fund would send them a set of fifteen Colportage library books. These were often used as part of the school library, or kept in classrooms for class use. The Book Fund had three standard sets: the first was for grades one and two, the second was for grades three through eight, and the last was for high schools. No complete listing of the books for these libraries survives, but each set probably included some hortatory works such as Dwight L. Moody's *Way to God*, as well as evangelically themed storybooks such as *Rosa's Quest*, or adventure stories such as *The Robber's Cave*, and *Tales of Adventure from the Old Book*. In addition to this library, the Book Fund delivered one specially edited version of the Gospel of John for each student. Students who memorized eleven Bible verses would receive a free *Pocket Treasury*, a short compendium of evangelically themed Bible passages and hymns. If they memorized twenty-eight gospel verses they received a free New Testament.[71]

Following the success of the southern mountain program, the BICA expanded the Bible distribution program to include other groups. School-age children in prison were prime targets, as were Catholic-school students in Louisiana and African-American public school students throughout the South. By the end of the 1920s, the "Pioneer" Book Fund had expanded to include public schools throughout the state of Wyoming. By 1929, 116 Wyoming public schools had been contacted, including over two thousand children in the Scripture-memorization program. The books that these young people received from the Missionary Book Funds of the BICA were all intended to convert them to fundamentalist Protestantism. All of the books were inexpensively produced, and most were written with an eye toward high-interest youth appeal. Within these parameters, however, there was considerable variety.[72]

The single most common book received by schoolchildren was a special edition of the Gospel of John. Editor T. C. Horton, an early leader of the Bible Institute of Los Angeles, began his evangelical effort in these pocket-sized Gospels with a short verse on the cover: "Here is a little book for you! / Just take it, now, and read it through. / Page sixty-six, verse thirty-one, / Believe it, and the work is done!" Eager young readers following this advice discovered chapter 20, verse 31: "But THESE [words] ARE WRITTEN, THAT YE MIGHT BELIEVE THAT JESUS IS THE CHRIST, THE SON OF GOD; AND THAT BELIEVING YE MIGHT HAVE LIFE THROUGH HIS NAME." Horton and the book fund missionaries who distributed this little book in immense numbers to Appalachian public schoolchildren were not satisfied to merely convert children to a generic Christianity. The conversion they were explicitly seeking was to the doctrines of fundamentalist Protestantism. The closing pages of the little book exhorted readers to accept the doctrines of the "Christian Fundamentals Association," including the inerrancy of the Bible, the deity of Jesus, Jesus's blood atonement for sin, Jesus's physical, bodily resurrection, and "His Personal, Bodily Return." In its stirring conclusion, it even offered this "Royal Resolution for Every Real Believer":

> I am living in the era of the World's crisis
> I am living in the era of the Church's crisis
>   THEREFORE
> . . . I will stand, by God's grace, with unquestioned confidence in the whole Word of God, and with the unsheathed Sword of the Spirit, contend for the faith once for all delivered, against all deceivers in school and church.

By distributing this inflammatory book so widely among Southern schoolchildren, the BICA sought to spread a specifically fundamentalist interpretation of Protestantism.[73]

Using this method, the Book Fund was able to deliver huge numbers of Bibles and literature to the public schools of Appalachia. In 1921, the "Mountain" Book Fund delivered 19,101 books to the region. By 1924, the total reached 47,103, and leaped to 271,214 by 1929. This number continued steady at about 250,000 per year through the 1960s. During the 1920s, they ultimately delivered approximately three-quarters of a million Bibles and religious books into the public schools of the Appalachian and Ozark regions alone.[74]

This immense distribution program largely escaped controversy, even in the controversy-filled 1920s. It managed to do so mainly by delivering Bibles and literature to regions in which the message was widely shared. Unlike in many larger cities, where politically savvy and influential minorities were able to exert influence on the school board, fundamentalist Bible missionaries such as Elizabeth Evans and the BICA took advantage of both the cultural homogeneity of many rural public schools and the students' desperation for reading material.

Although at least as successful as the antievolution campaign, this Bible crusade never became widely perceived as a peculiarly fundamentalist issue. Throughout the 1920s, nonfundamentalist conservative Protestants and many liberal Protestants worked just as diligently as fundamentalists to pass mandatory-reading laws and to deliver Bibles directly into schools. Unlike the evolution issue, these nonfundamentalists were never forced to either accept a fundamentalist label or abandon the campaign. Similarly, early fundamentalists who felt uneasy with popular new definitions of the fundamentalist movement could continue to push for Bible-reading in schools without having to embrace the new stereotype of fundamentalism.

In many cases, the remarkable aspect of both the direct-action campaigns and the drives for mandatory-reading laws was the lack of controversy surrounding them. Although the laws passed only over vociferous opposition in many places, they succeeded without a murmur of contention elsewhere. In most cases, the drive to ensure that public-school students would hear the Bible during their school day did not divide Protestants. This lack of controversy allowed fundamentalists and nonfundamentalists to work side by side. Their motives may have differed, but their goal was the same. This campaign, with no nationwide spotlight such as had focused on the Scopes trial in Tennessee, quietly did as much to change the nature of the nation's public school as the more ballyhooed antievolution crusade.

PART IV

Fundamentalism Transformed

CHAPTER 9

Fundamentalists and the New Fundamentalism

Hostile efforts to restrict the boundaries of the fundamentalist movement would not have had as much success had they been utterly rejected by fundamentalists themselves. Instead, in the second half of the 1920s, some fundamentalists embraced the new stripped-down understanding of fundamentalism. Many adopted a prickly defensive stance, one that proudly accepted the insults of liberals as badges of honor. They often fiercely refused to accept attributions of ignorance or bigotry. However, they enthusiastically affirmed the positive implications of hostile accusations. For instance, as liberals attacked fundamentalism as isolated, ignorant, primitive, and anti-intellectual, some fundamentalists defensively affirmed fundamentalism as part of a Southern, populist, anti-intellectual revival tradition. Furthermore, many prominent fundamentalists unintentionally reinforced the image of fundamentalist bigotry with their aggressive Protestant militancy. These efforts by fundamentalists, whether deliberate or unintentional, bolstered the new public image of the movement.

Even among the leaders who embraced this vision of fundamentalism, however, there remained significant differences of opinion about its exact meaning. As we have seen in Chapter 6, some fundamentalists, such as George Washburn of the Bible Crusaders and Edward Young Clarke of the Supreme Kingdom, vigorously promoted a conspiratorial, bigoted, anti-intellectual ideology. Others, such as George McCready Price, William Bell Riley and Bob Jones Sr., fought bitterly to defend the intellectual integrity of fundamentalism. Yet even among these three, each differed in his emphasis. Riley held out greater hope of retaining an image of fundamentalism as a movement with mainstream intellectual respectability. In contrast, Jones worked to establish an independent intellectualism, one that honored Western intellectual and artistic traditions while placing God and fundamentalist theology in the foreground. Price, along with other leading fundamentalist scientists, angrily defended his scientific legitimacy, while slowly accepting the need to build independent scientific institutions free from the hostile presuppositions of mainstream American science. Still other prominent leaders, including J. Frank Norris, paid mere lip service to the intellectual traditions of fundamentalism while confirming with their violent rhetoric and actions the accusations of liberals.

In spite of these differences, fundamentalists who defensively embraced restricted boundaries for the fundamentalist movement shared a strategy that included some degree of disengagement from mainstream American culture. Although they generally maintained an earnest hope of eventual cultural vindication, these defensive fundamentalists often embraced their continuing identification with fundamentalism as a sign of their continuing righteousness in the face of a morally decrepit mainstream culture. Sometimes reluctantly, they turned away from a mainstream culture that used "fundamentalism" as a term of opprobrium and built schools and a subculture that embraced those intended insults as marks of legitimacy.

One of the most persistent accusations of liberals was that fundamentalism was merely a regional movement. As we have seen, many nonfundamentalist newspaper reporters assumed at the Dayton trial that fundamentalism attracted only isolated hill farmers from Southern Appalachia. H. L. Mencken worked diligently to promote this stereotype. On his return from the Scopes trial, Mencken wrote with alarm that "in the rural sections of the Middle West and everywhere in the South save a few walled towns—the evangelical sects plunge into an abyss of malignant imbecility, and declare a holy war upon every decency that civilized men cherish."[1] Of course, fundamentalism claimed a much wider geographical base. Its leaders and leading institutions were scattered among cities such as Chicago, Minneapolis, New York, Los Angeles, Seattle, Toronto, and Fort Worth. Nevertheless, many leading fundamentalists embraced the limited regional stereotype of the movement, especially in the wake of the Scopes trial. John Roach Straton, in spite of the fact that he had built an successful career as a fundamentalist activist in New York City, identified "a rising tide of spiritual earnestness preparing in the South and West which will sweep the country" in the wake of the Scopes trial.[2] Gerald Winrod, himself a leader in the fundamentalist West, gave credit to the South for leading fundamentalist campaigns. In 1927, Winrod argued that the South should be the starting point for future fundamentalist activism, "because the South has led in the movement against modernism."[3]

Other leading fundamentalists had long argued for the regional nature of their movement. Since the first years of the 1920s, Texan J. Frank Norris had bemoaned the fact that northern infidelity was moving south. He warned readers that "our Southland, the boasted home of orthodoxy," was under threat from liberal Protestant theology.[4] Evangelist and school founder Bob Jones Sr., made similar heated regional arguments. In one sermon Jones delivered widely after the sensationalism of the Scopes trial, he promised his audiences, "We are not going to deliver the South to this rationalistic, atheistic leadership."[5] In the battles over fundamentalism and evolution in North Carolina, local fundamentalists repeatedly accused liberals of bringing in "foreigners" from the North.[6] Indeed, many North Carolina fundamentalists accused the public university, led by the "damn Yankee" president Harry Chase,[7] for supporting "modernists, Darwinian apologists, and Northerners."[8]

Part of this defensive embrace of regionalism included a new emphasis on the power of local politics. Especially in the wake of bruising publicity from national newspapers at the Scopes trial, many fundamentalists argued that local majorities could still be relied upon to support fundamentalist school policy. In some cases, this strategy had existed prior to the Scopes trial. T. T. Martin, for instance, argued in 1923 that local school boards were "absolute sovereigns; they can put in or out whatever teacher they will; no power on earth can force teachers on them." Martin suggested that these local school boards should be targeted first. If fundamentalists could convince school boards to fire or discipline those who taught evolution, Martin argued, students would be safe "until we can elect legislatures that will cut off all appropriations wherever Evolution is taught."[9] Other prominent activists had long agreed that the local approach was best. William Jennings Bryan, for instance, had urged the mayor of Chicago in 1923 to look into that city's policy on science education.[10] Other leading fundamentalists only embraced a local approach in the aftermath of the Scopes trial. In early 1926, Riley concluded that while state laws were good, securing a "Fundamentalist Committee in every city" to infiltrate school boards and root out teachers of evolution might be better.[11] Supreme Kingdom leader Edward Young Clarke even claimed to oppose state laws. Clarke insisted on a pure local majoritarianism, confident, no doubt, that majorities would oppose evolution. "You cannot legislate religion into the people," Clarke declared. "The question of teaching evolution . . . is before the people for decision and the responsibility for carrying out the peoples' choice rests upon the managements of the schools."[12]

Although, as we will see in Chapter 10, some fundamentalists tried to assert a more modern, "progressive" definition for fundamentalism, other leading fundamentalists attacked modern intellectual and educational developments.

These fundamentalist voices agreed with hostile liberals that fundamentalism necessarily included a kind of militant cultural traditionalism. They argued for a fundamentalist movement rooted in the past, before dangerous intellectual and cultural innovations had turned mainstream American culture away from the tenets of conservative evangelical Protestantism. Bob Jones Sr., for instance, never accepted what he lambasted as a "false educational philosophy." This philosophy, according to Jones, was rooted in the modern, progressive "self-expression, behavioristic idea of education." In order to save schools and souls, Jones insisted on returning "to the old time Puritan conception."[13] Jones was not alone in attacking modern ideas of childhood and education. Frank Gaebelein, founder of the fundamentalist Stony Brook School for Boys and son of prophecy writer Arno Gaebelein, agreed that the "behavioristic or mechanistic psychology . . . that underlies so many of the new methods of teaching" had put students' souls in desperate peril.[14] Norris had long attacked modern trends in education. He scorned modern intellectuals who had studied at elite schools only to lose their traditional evangelical faith.[15] In the wake of the Scopes trial, fundamentalists from Kansas to New York City harped on the refrain that America needed a "nationwide revival of old-fashioned Holy Ghost religion"[16] or "old fashioned gospel truths."[17]

For both liberals and fundamentalists, Fort Worth Baptist leader Norris personified the newly restricted image of fundamentalism. Norris's energetic embrace of the rapidly constricting boundaries of fundamentalism in the wake of the Scopes trial can be seen in his efforts to identify himself ever more closely with the label. As we have seen in Chapter 2, he had at first considered fundamentalism to be a purely Northern movement.[18] However, he soon made a defense of fundamentalism as such the focus of his activism. In 1927, he challenged the leadership of fellow fundamentalist leader Riley when he changed the name of his magazine from *The Searchlight* to *The Fundamentalist*. Like Riley, Norris hoped to identify himself as the leader of the fundamentalist movement. After a fight with Riley, Norris agreed to change the name again, but continued to identify the magazine as an organ of the fundamentalist movement, by calling it subsequently *The Baptist Fundamentalist of Texas*, and *The Fundamentalist of Texas*.[19] Norris also encouraged Northern reporters to label Norris with traditional barnstorming evangelical nicknames. Norris embraced these nicknames, including "Tornado" Norris, the "Texas Tornado," and the "Texas Cyclone."[20]

Although Norris's activism and rhetoric before and after the Scopes trial promoted the newly restricted image of the fundamentalist movement, Norris occasionally protested against accusations of bigotry or venality. As many other fundamentalists had, Norris took issue with liberals who called fundamentalists "ignorant," "bigots," and "cranks." He testily pointed out to a hostile editor that his Fort Worth hometown did not constitute the "backwoods."[21] He also protested against the mainstream press' prejudice against fundamentalism. In one article, he warned readers that "it is a well known fact that you cannot believe much that is said in the daily press concerning the ministry."[22]

Nevertheless, he repeatedly engaged in exactly the sort of militant, aggressive rhetoric that nonfundamentalist journalists had seized on as the essence of fundamentalism. For instance, in an introduction to a 1928 sermon by fundamentalist John R. Rice, Norris agreed that fundamentalism meant more than conservative evangelical theology. It also included a measure of militancy in the embrace of that theology. As Rice had argued, "Fundamentalism is not only what you believe but how strong you believe it. . . . It means, if necessary, offending and grieving people and institutions that have meant a great deal in my life."[23] Norris certainly did not shrink from offending those close to him. In fact, his vitriolic attacks on Baylor University earned him an unofficial censure at the hands of the Texas Baptist Convention.[24] This censure, however, only encouraged Norris to emphasize his militant rhetoric. He repeatedly described the controversies in the schools and Baptist denomination as a "war." He once argued, for instance, that "Fundamentalists should declare war on all fronts."[25] Further, he declared that the school fight was "a war to a finish. There will be no compromise."[26] Norris proudly told a *New York Tribune* reporter, "We fundamentalists . . . are determined to wage a relentless war against every phase of modernism. We propose to carry this war into every college and university and into every legislature." While liberals and nonfundamentalists sought to restrict fundamentalism to its stereotype as an overly aggressive, militant movement, Norris eagerly embraced and promoted a similar understanding of fundamentalism.[27]

Norris also worked to support other fundamentalists who similarly embraced the new image of fundamentalism. He never wavered in his support for Toronto's T. T. Shields, even when Shields's aggressive, acerbic rhetoric led to his ejection from his provincial Baptist convention.[28] Shields soon accepted a job as leader of the experimental fundamentalist university Des Moines University, with disastrous results, as described in Chapter 7. In spite of the bitter fruit reaped by Shields's aggressive and militant activism, Norris never flagged in his support.[29]

In addition, Norris embraced other reactionary activists, such as his local Ku Klux Klan. Many liberals and nonfundamentalists assumed that the fundamentalist movement and the Klan were identical. As we have seen, the two

movements emphasized different aspects of Protestantism and cultural traditionalism. Nevertheless, Norris again confirmed the stereotyped image of fundamentalism with his trenchant support for the Klan's anti-Catholic bigotry.
Norris had long supported such vigilantism. In 1917, he praised a group of
female nightriders in Pittsburgh who had burned down a brothel. "It would be
a blessing to the city of Fort Worth," Norris wrote, "if we would be visited some
night by the same band of Women."[30] His support for such violent activism continued into the 1920s. He often published advertisements for local Klan groups
in his newspapers.[31] He also supported candidates for political office based on
their Klan affiliation.[32] He even sponsored a Klan minstrel show at the First
Baptist auditorium.[33] Norris's ardent support for the Klan may have grown out
of his violent anti-Catholicism. Norris repeatedly denounced the conspiratorial
military organization of the Catholic Church and the Knights of Columbus.
He warned his readers that Catholics yearned to "burn to ashes every Protestant
church and dynamite every Protestant school."[34] He even denounced a professor at his local Baylor University on the charge that the professor was Catholic.
This paranoid, violent bigotry against Catholics helped cement the new image
of fundamentalism as a whole. While many prominent fundamentalists had
long disavowed the vigilantism, bigotry, and secrecy of the Klan, Norris enthusiastically embraced it as part of his definition of fundamentalism.[35]

More than such violent, aggressive rhetoric, however, Norris's deadly encounter with a Fort Worth rival may have contributed the most to the new image of
fundamentalism as an extremist, violent, antimodern Southern movement. The
rival, D. E. Chipps, had sustained a long public quarrel with Norris. Chipps, a
Fort Worth lumber dealer, had objected to Norris's repeated insults against many
prominent citizens of Fort Worth. On July 14, 1926, Chipps publicly threatened Norris and followed Norris back to Norris's office for another angry ver-
bal exchange. A few days later, according to Norris, Chipps called Norris on the
phone, then burst into Norris's office brandishing a gun. According to Norris, the
two men struggled, and the gun went off, mortally wounding Chipps. In Norris's
words, the murder was entirely "self defense. He threatened to kill me. I had to do
it." A jury agreed, based mostly on Norris's testimony and that of a single witness.
Norris's popularity continued unfazed, but this image of a primitive Southern
blood feud contributed to the new stereotype of fundamentalism.[36]

Other prominent fundamentalist leaders worked quixotically to maintain a
wider definition for fundamentalism, beyond this backward-looking, violent, Southern antimodernism. Riley had committed himself and his career to

fundamentalism since the earliest days of the movement. Although he had been raised in Kentucky, Riley built his successful career as a Minnesota Baptist with aspirations to national and international leadership of an intellectually respectable fundamentalist movement. As such, Riley could not support the image of fundamentalism as merely Southern anti-intellectualism. Although by the end of his life in 1947 Riley had accepted a bigoted, conspiratorial explanation for the ebbing fortunes of fundamentalism in the late 1920s, in the years immediately following the Scopes trial he still hoped to single-handedly force open the boundaries of fundamentalism to include room for his Northern, urban, intellectually respectable definition of the movement.

In the aftermath of the Scopes trial, Riley worked diligently to counter the image of fundamentalism as a bigoted, anti-intellectual movement. He avidly read liberal and mainstream newspapers and magazines, and was keenly aware of the growing popularity of negative, stereotyped ideas about the meanings of fundamentalism. As he began his campaign for a fundamentalist school law in Minnesota, for instance, he knew that liberal enemies blithely assumed that Riley was fighting for a "medieval" Minnesota, not a "modern" one.[37] He argued, on the contrary, that he was the only one fighting for open-minded discussion of the issues. In his attack on the University of Minnesota's proevolution policy, he blasted university leaders as exhibiting a "contemptible piece of prejudice" in their antifundamentalist attitude.[38] He accused university administrators and professors of bullying students into conformity with liberal ideas. "Every student," Riley charged, "if he do any independent thinking . . . is yet cowed into acquiescence."[39]

Riley worked hard in the aftermath of the Scopes trial to project a public image of himself as an erudite, sophisticated theologian bemused by popular misunderstanding of the finer points of fundamentalism. At one point during the controversy over the Minnesota antievolution law, Riley sought to portray himself as morally superior to petty attacks on the image of the fundamentalist movement. "Every time I hear the argument that this is a controversy between experts on the one hand, and, as someone has said, 'organized ignorance,' on the other, I smile," he assured the St. Paul *Pioneer Press*. "This is not a debate between the educated and the uneducated."[40] But many Minnesotans, like many Americans nationwide, believed it was such a debate. Recall that in his November 1926 appearance at the University of Minnesota, a student prankster had lowered a live monkey on the stage. On that occasion as well, Riley adopted a pose as condescendingly amused.[41] However, to maintain a wider definition of fundamentalism, Riley's public "smile" had to mask the strain of fighting a rearguard action. In one article, Riley spelled out the nine tenets of the World's Christian Fundamentals Association creed. Then he bitterly pointed out that those beliefs did not include belief in "'a flat earth,' . . . 'an immovable world,' . . . [or] 'a canopy of roof overhead.'" Riley knew, however, that many Americans in the wake of the Scopes trial thought those antiquated beliefs signified exactly the nature of fundamentalism. Riley spent much of his mental and physical energy refuting charges of ignorance and exaggerations of the story of the Scopes trial. In a 1927 article, Riley protested, "This charge of ignorance in realms of science against the leaders of fundamentalism has about as much basis of truth as had the statement from the university professor that the author of the Tennessee antievolution bill had, upon learning that the Bible was not made in heaven and dropped down, expressed his regret that he ever wrote or advocated the passing of the bill."[42] When he could, Riley attempted to turn antifundamentalist news into fundamentalist success. For instance, he gleefully thanked liberal writer Harbor Allen for the free publicity when Allen damned the creation of new fundamentalist groups in the wake of the Scopes trial. Even so, Riley recognized that such hostile reports had become the norm in the mainstream press.[43]

As he felt the struggle over the mainstream popular image of fundamentalism slip away from him, Riley doubled his efforts to control at least the fundamentalist side of the battle. He may not have been able to pressure liberals to change their ideas, but he could exert pressure on fellow fundamentalists to conform to Riley's vision of fundamentalism. In early 1926, Riley urged all local antievolution organizations to subordinate themselves to his World's Christian Fundamentals Association (WCFA). Riley suggested that smaller groups could apply to become local arms of the WCFA. This central control of fundamentalist organizations would have allowed Riley to assert much greater control over the ways fundamentalism was promoted.[44]

Riley also demonstrated a sharpening insistence on doctrinal purity. Riley and his WCFA had always been committed to the doctrine of a premillennial return of Jesus. That is, Riley interpreted scripture to mean that Christ would return after a period of tribulation to usher in a thousand-year earthly reign of peace and harmony. Most early-twentieth-century fundamentalists shared this belief, but not all. In the first years of the 1920s, Riley was friendly with those, such as William Jennings Bryan, who believed in a postmillennial return of Christ. Riley even offered Bryan a leadership role in the WCFA, in spite of this difference in doctrine.[45] In 1922, Riley had insisted that his movement was open to nonpremillennialists.[46] After the Scopes trial, however, Riley began to insist more vehemently on such doctrinal issues as a test of fundamentalist credentials. He mocked "Funny Fundamentalists," who may have sided with fundamentalists on many issues but did not ascribe to premillennial theology.[47] He insisted that the timing of Christ's return was a question that had been "forever settled in heaven." It was part of God's truth, Riley insisted in 1927, "As unchangeable as imperishable."[48]

Riley insisted on his leadership of the fundamentalist movement so adamantly partly because of his belief in himself as a natural leader, but also because he knew of the difficulties of building fundamentalist consensus about the proper definitional boundaries of their movement. He attacked fellow fundamentalist Norris in 1927 when Norris retitled his magazine *The Fundamentalist*. This battle was more than a clash of overinflated egos. It became by 1927 a clash over the right to define the boundaries of fundamentalism from the inside. Riley needed to maintain control over the term in order to successfully maintain a wider identity. Norris, the shoot-'em-up "Texas Cyclone," embodied the stereotype against which Riley was unsuccessfully fighting. Although they worked closely together in the early years of the decade, once pressure mounted to restrict fundamentalism's image to a violent, rural, anti-intellectual stereotype, the two clashed. Norris embraced the new stereotype, and Norris's embrace fueled Riley's vindictive attack. Riley did more than fight with Norris over the rights to the fundamentalist name. Riley's bitterness over his failure to control the boundaries of fundamentalism led to a long-lasting grudge against Norris, his former close ally. Even in 1938, Riley continued to attack Norris as a charlatan engaged in "dastardly" "religious racketeering."[49]

Unfortunately for Riley, his quest to unilaterally assert the definition of fundamentalism in the wake of the Scopes trial was doomed from the start. One committed fundamentalist, no matter how influential, could not single-handedly define boundaries for fundamentalism in the years after the Scopes trial. This was especially true when several of Riley's fellow fundamentalists had embraced the stereotype of fundamentalism, in spite of Riley's protests. Instead, the image of the fundamentalist movement formed in the wake of the Scopes trial as both fundamentalists and outsiders implicitly moved toward a consensus about the nature and boundaries of the movement. In Riley's case, he unsuccessfully campaigned to maintain a wider vision of a fundamentalism that even nonfundamentalists could find intellectually respectable.

Nevertheless, Riley clung to his own definition of fundamentalism and built a significant regional "empire" in Minnesota. He was largely able to control the meanings of fundamentalism in his region and used his influence to construct a powerful denominational and political voting bloc throughout the state. However, Riley found that his determined defiance of fundamentalism's new image could not change popular opinion beyond his regional power base. He struggled in the decades following the Scopes trial to understand why his prominent leadership of the movement did not translate into the ability to define the boundaries of that movement. In the end, Riley adopted a conspiracy theory to explain his frustrated struggle. According to Riley, it was a sinister cabal of "Jewish Communists" that had derailed his heroic efforts to maintain a fundamentalist movement that nonfundamentalists would find intellectually respectable.[50]

Bob Jones Sr. embraced fundamentalism in a different way in the wake of the Scopes trial. As a Southerner, Jones was comfortable with regional stereotypes of the movement. As a successful itinerant evangelist, he had long embraced a self-image as a crusader against the ills of the modern world. Although he resented accusations of ignorance and took pains to refute them, Jones energetically supported other elements of fundamentalism's new image. Jones used the new image to build a powerful educational network in the years following the Scopes trial.

Although Jones toured the entire country as part of his evangelical career in the early 1900s, he often argued for the spiritual superiority of the South. And, although he claimed to be attracted to the original Florida location of his new university in 1926 because it could attract students from the North, the new college relied on students from the South to get started.[51] More telling, Jones promoted an image of fundamentalism as part of a rural Southern cultural traditionalism. As had liberal opponents, Jones asserted that fundamentalism implied a rural, antimodern militancy, one that looked backward in time. The answer to America's moral, spiritual, and cultural ills, according to Jones, was in a fundamentalist movement defending "old-time" Protestantism. One way to solve the dilemmas of modern society, Jones argued, was to "bring back to the schools of this nation the Word of God and the old time religion."[52] The problem with American society in the 1920s, Jones preached, was not an excess of materialism, but the fact that "we have given up our old time country idea of God and decency. . . . Degeneracy has already set in." Even the physical safety of Americans was at risk in modern America, according to Jones, due to "Paved highways, automobiles, and modern travel." These rural, regional, and backward-looking assertions accorded exactly with the image promoted by liberal opponents of fundamentalism.[53]

Jones used this definition of fundamentalism to attract considerable support among like-minded fundamentalists. For instance, in the years following the Scopes trial, Jones promoted his new college as a "continuous Chautauqua."[54] By this time Jones made this promise in 1928, the traveling-lecture format of chautauquas had already become a symbol of old-fashioned, rural Victorianism.[55] Jones used the image proudly to associate his new school with this educational institution symbolic of an older preurban America. Many fundamentalists found this association compelling. As we have seen in Chapter 7, Bob Jones College quickly attracted hundreds of students. Although mainstream America may have moved away from the chautauqua as the ideal educational model, these fundamentalist students and their families apparently agreed with Jones about the desirability of such a nostalgic educational model.

Jones' embrace of the new image of fundamentalism did not mean that he accepted liberal accusations of fundamentalist ignorance and anti-intellectualism. Jones often recalled in later years that he had founded his college, in part, to dispel the popular association of fundamentalism with ignorance in the wake of the Scopes trial. He remained throughout the 1920s extremely touchy about liberal accusations of ignorance.[56] The first dean of Bob Jones College (BJC), W. E. Patterson, examined BJC students' standardized intelligence test results in 1928. The scores of students at "several colleges of high standing," Dean Patterson claimed, averaged 150. The average score for new students at BJC, in contrast, was 158.55. Patterson concluded triumphantly that these results proved that fundamentalism did not imply a lack of intelligence.[57]

This proud intellectualism, however, did not mean that Bob Jones had accepted the skeptical attitude central to modern "'so-called' science."[58] For instance, Jones proudly refuted some liberal stereotypes by advertising the teaching of evolution at BJC. Jones claimed to teach not just Darwin, but even Marx as well. The difference between his school and mainstream public colleges, Jones explained, resulted from the fact that "we teach it as a theory and not as a fact." In this way, according to Jones, students would receive all the benefits of an excellent academic education, without forgetting that "the spiritual is more important than education." He warned students, "There are some things that education alone cannot do. Nothing but God in the soul through the miracle of regeneration can save the individual and make the type of character that will save our civilization." Nevertheless, Jones insisted that every student at his school would receive "the best educational advantages."[59]

Other fundamentalists agreed that a continuing embrace of fundamentalism did not necessarily imply an abandonment of intellectual rigor. In fact, in the years following the Scopes trial, leading fundamentalist scientists continued to battle for recognition of their scientific achievements. However, as popular stereotypes about the image of fundamentalism gained currency, even leading fundamentalist scientists such as George McCready Price, Arthur I. Brown, and Harry Rimmer implicitly conceded the field of mainstream American science to their evolutionist rivals. Although they never abandoned their claims to scientific legitimacy and even superiority, they quietly moved away from attempts to convince nonfundamentalist audiences of their claims to legitimacy. Instead, as had Riley and Bob Jones Sr., these scientists eventually recognized the need to build independent institutions, ones not held to the hostile standards of nonfundamentalist scientists.

Several new groups sprouted in the immediate wake of the trial to fight for the superiority of fundamentalist science. For instance, Los Angeles-based evangelist Robert "Fighting Bob" Schuler joined with Riley to form the short-lived Defenders of Science versus Speculation. This ephemeral organization never achieved its ambitious goals, but its name alone testified to the belief of many fundamentalists that their understanding of science could be successfully defended against the claims of mainstream science.[60] And in the years following the trial, other new fundamentalist organizations continued to fight for the superiority of their traditional view of science. The Bible Crusaders, for instance, promised local fundamentalists that the Crusaders could supply "scientists and lecturers of national reputation."[61]

In addition, the many attacks upon the image of the fundamentalist movement served to increase the prestige among fundamentalists of the small cadre of evangelists with mainstream scientific credentials. Arthur I. Brown, for instance, found himself recruited by fundamentalist groups such as the WCFA. As historian Ronald L. Numbers has noted, Brown found himself besieged in the years following the trial by pleas from fundamentalists seeking his full-time help. In November, 1925, Brown agreed to a one-year leave from his medical profession, "at the urgent solicitation of many prominent Fundamentalist leaders." Due to the insatiable demand for his scientific presentations about the fallacies of evolution, Brown was able to permanently, and profitably, abandon his medical practice for a busy schedule of speaking engagements across the country. Advance publicity for these engagements consistently emphasized Brown's mainstream scientific credentials. Brown was an M.D. with a British surgery degree. These credentials continued to be of primary importance among fundamentalists. Gerald Winrod, leader of the upstart Kansas Defenders of the Christian Faith, urged his readers to "Notice the degrees attached to the writer's name . . . realize that he was trained in some of the best Universities in Europe, and you will know why he is recognized as an authority." Other hosts described Brown as the "greatest scientist in all the world." Clearly, these fundamentalist audiences had not abandoned their belief in the superiority of their scientific experts to those of the scoffing scientific mainstream.[62]

In spite of these fundamentalist accolades, Brown himself slowly accepted the need to abandon his quest for recognition among mainstream scientists.

In his writings before the Scopes trial, Brown had repeatedly asserted his alliance with "leading scientists." Those scientists, Brown contended, agreed with Brown about the scientific illegitimacy of evolution. Even before the Scopes trial, Brown recognized the enmity of those who claimed "'the consensus of scholarship' has accepted evolution," but in those earlier years Brown argued that "many great scientists . . . have utterly repudiated this idea."[63] By 1927, however, Brown continued to claim status as a scientist but he conceded that he had been cast out of the scientific establishment. He boasted that he would continue his antievolution activism, aware that he would "risk the penalty of excommunication from the ranks of the 'intelligentia,' [sic] and consignment to the bottomless depths of that region of darkness inhabited by ignorant and obscurantist opponents of the 'law of evolution.'"[64] Like other prominent fundamentalist scientists, Brown slowly came to a realization of the need to establish a separate scientific establishment, one immune from the "scorn" of America's mainstream scientists.[65] This independent establishment, Brown hoped, could explore a productive research agenda, unburdened by the destructive secular assumptions of most mainstream American scientists. Those scientists, Brown accused, had become merely "superficial thinkers [or] servile followers of a blatant pseudo-science."[66]

Harry L. Rimmer, another leading voice for fundamentalist science, also continued to fight against the scientific legitimacy of evolution after the Scopes trial. Rimmer claimed some formal training in medicine, although he had not completed his MD. Nevertheless, Rimmer maintained an exhausting schedule in the years following the Scopes trial, traveling across the country and debating evolution supporters in a variety of venues. In all his debates, Rimmer presented himself as a scientific expert. He customarily added the honorary titles "Doctor of Science" and "Doctor of Divinity" to his name, the first bestowed by Wheaton College, and the second by Colquith College. Other leading fundamentalists eagerly recognized Rimmer's mainstream scientific pretensions. Riley claimed Rimmer had "made many important discoveries in physics, chemistry, and biology." Many nonfundamentalists, however, were unimpressed. One hostile listener claimed, "As a scientist, Rimmer is a joke." This lack of mainstream credibility did not intimidate Rimmer or many in his fundamentalist audiences.[67]

Like many fundamentalists who continued to embrace fundamentalism in the years following the Scopes trial, Rimmer fought gamely to turn the tables on his enemies. Rimmer agreed with Brown that the voices of evolution had come to represent "intrenched [sic] bigotry." Rimmer lamented the fact that when he continued against "this fortified error," he was "called ignorant and an enemy of science!" Rimmer's complaint, however, demonstrated the building consensus about the nature of fundamentalism. Rimmer, like Brown, was confronted with a choice between abandoning all claims to scientific legitimacy, or building a separate scientific establishment that recognized his credentials. Although he continued to fight with mainstream scientists about the scientific legitimacy of fundamentalism throughout the 1920s, by the early 1940s Rimmer had revived his dormant Research Science Bureau, in order to build a separate fundamentalist establishment. This research bureau, originally founded by Rimmer in 1921, would "encourage and promote research in such sciences as have direct bearing on the question of the inspiration and infallible nature of the Holy Bible."[68]

Another fundamentalist scientist who wrestled with the shrinking boundaries of fundamentalism was George McCready Price. For Price, the Scopes trial had come as the last straw in a long struggle with the mainstream American science establishment. To a greater degree than either Brown or Rimmer, Price had fought for recognition from that establishment since the beginning of the 1920s. By the time of the trial, Price had already become embittered by two acrimonious disputes with James M. Cattell, editor of *Science* magazine. In each case, Cattell had published vicious attacks on Price's scientific credibility. Further, Cattell refused to publish Price's rebuttals to either attack. Worst of all, Cattell's dismissal explicitly rejected Price's claim to the mantle of science. *Science* would not print Price's defenses, Cattell wrote, because they "would not be of interest to scientific men."[69]

As a veteran of these humiliating conflicts, Price may have had a better sense of the probable outcome of the trial than Bryan and other leading fundamentalists. When Bryan pressed Price in the days leading to the trial to come lead their team of scientific experts, Price sought to dissuade Bryan from making the trial a contest between scientific experts. The trial, Price believed, was "not a time to argue about the scientific or unscientific character of evolution theory." Price understood better than Bryan how far removed his scientific expertise was from that of the American mainstream.[70]

After the trial, Price personified the analysis historian Joel A. Carpenter has offered about the fundamentalist movement as a whole during the 1930s. "In retreat from public embarrassment," Carpenter has suggested, "fundamentalists cultivated distinctive religious communities . . . fundamentalists weathered their defeats and humiliation and not only survived but thrived."[71] Unlike Brown's and Rimmer's continuing eagerness for public controversy and debate in spite of their growing recognition of mainstream obstinance, Price retreated from publicly challenging mainstream scientists after the trial. His last public debate took place in the immediate aftermath of the Scopes trial. At that debate, held in London against rationalist philosopher Joseph McCabe, Price continued to maintain the scientific superiority of his position. The theory of evolution, Price announced, may have been satisfactory, for the times of comparative ignorance of the real facts of heredity and variation and of the facts of geology which prevailed during the latter part of the nineteenth century; but that this theory is now entirely out of date, and hopelessly inadequate for us. . . . We are making scientific history very fast these days; and the specialist in some corner of science who keeps on humming a little tune to himself, quietly ignoring all this modern evidence against Evolution, is simply living in a fools' paradise. He will soon be so far behind that he will wake up some fine morning and find that he needs an introduction to the modern scientific world.

Price unsuccessfully sought, as had Brown and Rimmer, to convince nonfundamentalist audiences in the wake of the Scopes trial that antievolutionary science was superior to evolutionary science on the terms of mainstream scientists themselves.[72]

By September of 1925, however, when this London debate took place, the popular consensus against the scientific legitimacy of fundamentalism and its leading scientific experts had become so powerful that the audience heckled Price unmercifully. "Do not confine your reading wholly to one side," Price pleaded in response to one scornful outburst from the audience. "How can you know anything about a certain subject if you read only one side of the case? There is plenty of evidence on the other side, and this evidence is gradually coming out." After this debate, Price left the stage feeling humiliated, and he never engaged in another public debate.[73]

This humiliation did not mean the end of his career by any means. He wrote and published tirelessly to promote his scientific views among fundamentalists and other conservative Christians. He also refused to embrace the disparaging humor of fundamentalism's enemies. In 1927, he turned down an invitation to edit a Seventh-day Adventist antievolution journal, partly due to its undignified title, "the Monkey Magazine." Instead, Price insisted, he would continue to formulate "a dignified, scholarly presentation of facts and arguments."[74] Price would eventually realize his ambition. By the late 1930s, Price and other fundamentalists had recognized the need for an entirely independent scientific community. To help realize this goal, Price founded the Deluge Geology Society. He hoped to create a biblically based independent scientific community. Never again would Price or other fundamentalist scientists need to rely on mainstream scientific institutions for their sense of legitimacy. The educational controversies of the early 1920s, with their culmination following the Scopes trial, finally convinced Price of the need for these alternative cultural institutions.[75]

Other fundamentalist leaders who continued to embrace fundamentalism in the wake of the Scopes trial often faced similar dilemmas. The intense publicity of fundamentalist educational campaigns squeezed the definitions of fundamentalism in ways many fundamentalists found uncomfortable. Riley tried and failed to maintain a self-image as a fundamentalist free from the implications of ignorance and bigotry foisted upon it by liberals and nonfundamentalists. After an embittering battle, Riley retreated to a regional fundamentalist empire among Minnesota Baptists, where his influence allowed him to dominate the local image of fundamentalism. Bob Jones Sr. embraced liberal accusations of Southernism and cultural traditionalism. However, he insisted on a new understanding of fundamentalist intellectual rigor. As had Riley, Jones built his own educational island free from the constraining influences of hostile attributions about fundamentalist ignorance. Norris embraced the hostile stereotype of fundamentalism with only passing protests against charges of bigotry or ignorance.

Like fundamentalist scientists, these fundamentalists embraced the fundamentalist label in spite of increasingly influential attacks on their movement in the wake of the Scopes trial. As liberals and nonfundamentalists had asserted severely limited boundaries for fundamentalism, so many leading fundamentalists embraced those newly restricted boundaries. These fundamentalists turned some of the accusations of liberals into proud markers of their movement. In doing so, they helped to build a formidable new popular image about the meanings of fundamentalism, one rooted in regionalism, cultural traditionalism, and opposition to the intellectual primacy of skepticism.

CHAPTER 10

Fundamentalists Outside
the New Fundamentalism

In the fall of 1927, Thomas Gillespie took the stage to deliver the opening address at a Bible conference in Pittsburgh. Like many in his audience, Gillespie felt that the American norms and cultural values on which he had built a successful life and career had shifted out from under him. His aggressive, energetic leadership of one of the largest steel and iron firms in the region was still successful in the new age, but he nevertheless found himself under attack. Instead of his accustomed role as a respected businessman and pillar of the community, he squirmed awkwardly on the receiving end of public ridicule and attack. Why? Because of his sincere commitment to his conservative Presbyterian beliefs, which he thought reflected the best moral traditions of his country.

This exasperating situation led Gillespie to welcome the bully pulpit from which he expressed his frustration at the ignorance and maliciousness of his foes. He mocked those benighted liberals and secularists who lumped all Bible-believing Christians into the same mold. He knew that critics labeled fundamentalists as "stand patters," "traditionalists," or even "reactionaries." Gillespie marveled that liberals could blithely equate his urban, bourgeois, respectable fundamentalism with that of the Tennessee backwoods. He hoped to ward off these new attacks by reminding his Bible conference audience that "we claim to be *progressives*," and no amount of facile generalizations could shake fundamentalists such as Gillespie from that comforting self-knowledge.[1]

Gillespie articulated the frustration felt by many fundamentalists in the years following the Scopes trial. He was disheartened by the widespread popularity of the new image of the movement. Like many activists, he had been attracted to the fundamentalist movement in the early 1920s as a wide coalition of conservative evangelical Protestants. After the extraordinary publicity surrounding fundamentalist school campaigns, including the Scopes trial, he did not feel comfortable with the new public image of fundamentalism that only had room for rural, Southern, antimodern, anti-intellectual revivalism.

Like Gillespie, many leading fundamentalists sought to combat the definitions of fundamentalism that had gained influence as a result of fundamentalist educational activism. William Bell Riley sought earnestly to refute charges of ignorance and bigotry. He pointed to fundamentalism's long intellectual tradition. Riley, however, remained committed to the fundamentalist label no matter how its popularly accepted definition may have shifted. When he lost his fight to maintain wider definitional boundaries for the movement, he clung bitterly to fundamentalism and took solace in his ability to control his Minnesota educational empire. Other early fundamentalists did not have such a fierce commitment to fundamentalism. As they resisted the ascriptions of liberals, these less-committed fundamentalists also reminded audiences, as Gillespie did, of their claims to other cultural identities, such as progressivism.

In the end, however, their efforts to assert a more inclusive definition for the fundamentalist movement were unsuccessful and many fundamentalists unobtrusively abandoned the label. To extend sociologist of science Thomas Gieryn's cartographic metaphor, these fundamentalists quietly allowed the boundaries of fundamentalism to shrink past their own positions. That is, they remained in the same cultural space, along the boundaries of early fundamentalism. As those boundaries constricted, these early fundamentalists passively allowed themselves to be defined out of the newly restricted fundamentalist movement. They did not change their own fundamental beliefs about the Bible or education. Nor did they abandon their commitment to activism that had remained outside the shrunken boundaries, such as drives for mandatory Bible-reading laws. Many even continued to maintain a reluctant and ambivalent association with fundamentalism. But they no longer eagerly embraced public campaigns, such as the drive to ban evolution from schools that had become inextricably equated with fundamentalism, nor did they identify themselves as fundamentalists unless forced to by the absence of acceptable alternative labels.

It would not be until the late 1940s that a new evangelical identity emerged more in line with this ambivalent position. A new generation of evangelicals, those who had felt alienated by the restricted definitional boundaries of 1920s-era fundamentalism, embraced theologian Carl Henry's 1947 call for a "progressive Fundamentalism."[2] Henry and his peers succeeded where Gillespie and other 1920s fundamentalists had failed. The next generation was able to construct a neoevangelical identity that included fidelity to traditional interpretations of the Bible while still achieving mainstream intellectual respectability.[3]

For fundamentalists in the 1920s, however, that kind of wider definition for fundamentalism remained elusive. Even more frustrating, the complex struggle to define the fundamentalist movement went on regardless of participants' ability to articulate their own experiences. Most fundamentalists could not clearly define their complicated relationship to the fundamentalist label. They insisted on a fundamentalist interpretation of the Bible, for instance, but they shied away from identification with the post-Scopes image of ignorance and anti-intellectualism.

Several leading fundamentalist intellectuals, however, struggled to understand and to express their growing dissatisfaction with the shrinking boundaries of fundamentalism. Lewis Sperry Chafer, Bible scholar and founder of the Dallas Theological Seminary, worked to explicate his complicated relationship with fundamentalism. A similarly complex articulation of this quiet abandonment of the fundamentalist label came from Presbyterian scholar J. Gresham Machen. A third contemporary intellectual who managed to analyze the shifting patterns of fundamentalism was James M. Gray of the Moody Bible Institute. Each of these thinkers sought to define himself in relation to the rapidly shifting boundaries of fundamentalism. Most activists, however, expressed their dismay at the confusing nature of fundamentalism in terms more similar to Thomas Gillespie's anguished but inexact reminder that he claimed to be a "progressive."

Many fundamentalists in the 1920s had long held traditionally progressive views. James Gray of the Moody Bible Institute (MBI) had campaigned tirelessly in favor of equal pay, equal voting rights, and equal property protection for African Americans. For Gray, theology trumped social custom. Since all men and women were created from Adam and Eve, Gray reasoned, racial restrictions could not be justified.[4] Gray's antiracism was shared among many early fundamentalists. New York pastor and activist John Roach Straton led many public antiracism campaigns as part of his continuing crusade for public morality. He also worked for other traditionally progressive reforms. In *The Scarlet Stain on the City* (1916), he fulminated against "such evils as unjust wages, especially to women workers, child labor, and hell-black social evil, lawlessness, and the awful shame and disgrace of the liquor traffic!"[5] Riley also began his career as a leader of both "evangelism and reform."[6] When he first arrived in Minneapolis in 1897, Riley's first public campaigns fought for prohibition, as well as for the rights of workers. In 1906, he attacked the greed of "growing and grasping corporations" for their continued abuse of average working citizens.[7]

Later fundamentalists continued to attach their educational campaigns to the progressive-era drive to assert government control over big business. Agitation against monopolies and trusts had simmered since the late nineteenth century, and Congress took its first step toward regulating business in earnest in 1890, with the passage of the Sherman Anti-Trust Act. In 1906, Congress passed the Pure Food and Drug Act to regulate the quality and safety of consumable goods. Fundamentalist author George McPherson hoped to associate his campaign for the restriction of college curricula to these successful progressive crusades. "Our Government has undertaken to regulate business," McPherson argued, "but the hour has come to regulate our higher education. . . . We protect . . . our food from adulterations, and shall we not protect our young men and women, in our institutions of learning?"[8]

The adoption of successful progressive-era rhetoric was common among fundamentalists in the 1920s. Many progressive-era reformers had succeeded at passing laws that curbed child labor. Other laws established special courts for young offenders. To win support for these reforms, activists often compared the developmental danger children faced when exposed to adult institutions to physical danger.[9] Consider, for example, the argument by progressive-era reformer Louise Bowen. She argued that her organization, the Juvenile Protective Association, "only felt that something must be done to build a fence at the top of the cliff in order to cheat the ambulance, symbolized by the Juvenile Court, waiting at the bottom of the precipice to rescue the childish victims." It made sense to either protect children from falling under the influence of immoral urban influences or deal with the certain criminality later.[10]

The goals of fundamentalism's reforms during the 1920s differed, but fundamentalist activists used exactly the same rhetoric to win support. Evangelist William E. Biederwolf, for example, argued that "atheistic theories are being thrust into our college curriculums, into our literature and our public lectures, and into the textbooks of our public schools, and even taught to lisping children in Sunday-school. It is altogether a question of building a railing around the top of the cliff or running an ambulance at the bottom of it."[11] William Jennings Bryan used similar child-safety rhetoric. To counter the argument that teachers had an inalienable right to teach whatever they considered appropriate, Bryan thus responded: "If they leave it to each student to look out for himself, they are like the drivers who dash by a school in recess time, remarking that the public highway is to drive on and that children must look out for themselves." Although fundamentalists sought to ban evolution and other "atheistic theories" from schools, instead of the earlier goal of limiting child labor or establishing special courts of law, the language of the fundamentalist movement copied that of the earlier progressive movements.[12]

Many fundamentalists asserted that they were, in fact, in favor of such progressive-era innovations as "child-centered" education. One educator with the Moody Bible Institute of Chicago concluded, "We believe that instruction should be child-centered in a certain sense." The writer wanted education to focus on the needs and capacities of individual children, as progressive reformers routinely advocated. The only restriction on the new rule, in his opinion, was that the integrity of the Bible must not be compromised. School should be child-centered, in other words, as long as it did not distort the meaning of the Bible.[13] Bible institutes also often insisted on a reformist pedagogy for their Bible classes. James Gray's "synthetic" method was widely used, because it was seen as an improvement over traditional approaches.[14]

Other fundamentalists harped on their own progressivism in a defensive way. Author M. H. Duncan opened his fundamentalist tract *Modern Education at the Cross-Roads* with the assertion that the "author is a progressive in education." Duncan argued that the Bible itself had been maligned. In fact, Duncan asserted, "It is a very progressive book."[15] Other fundamentalist activists joined in this assertion of their own fundamental progressivism. Elizabeth Evans, the fundamentalist educational home missionary, remembered her campaign to unite New England fundamentalists as successful among "really progressive" individuals.[16]

This use of progressive-sounding language must be understood in the context of the 1920s, however. During that decade, the label "progressive" was used to sell everything from political reform to personal hygiene products. One contemporary advertisement for razor blades, for instance, urged buyers to "Be Progressive," and buy only one brand of blades. When 1920s-era fundamentalists used the language of "progressive" reform, they often only hoped to associate themselves with a vague feeling of success.[17]

Beyond simply paying lip service to progressivism, at least one leading fundamentalist articulated a vision of fundamentalism closely aligned with ideas generally considered "progressive." Bryan, after all, had built his long successful political career as "The Great Commoner." He had championed traditionally progressive ideas such as direct election of senators, a graduated income tax, and, most prominently, progressive change in America's monetary policy. For Bryan it was natural to continue to claim the label "progressive" to describe fundamentalist educational activism. He often linked his fundamentalist activism with his progressive reputation. He described his antievolution crusade as only the most recent populist "reform" of the progressive movement.[18] Upon his death, bluegrass musician Charlie Oakes penned a tribute that underscored the consistency of Bryan's lifelong populist message: "He fought the evolutionists, infidels, and fools / Who were trying to ruin the minds of children in the schools / Three times he ran for president, but the capitalists wouldn't let him win / Because he was a friend to the poor and to the working man."[19]

The question of what may have become of the fundamentalist movement if Bryan had not died in the immediate aftermath of the Scopes trial is difficult to ignore. After Bryan's unexpected demise, leading fundamentalists such as Riley struggled unsuccessfully to maintain a wider, nonbigoted, intellectual image for fundamentalism. Like Riley and Bryan, many fundamentalists had been wounded to the quick by the popularly accepted accusations of ignorance and bigotry at the Scopes trial. Many fundamentalists regarded themselves as favoring tolerance and open-mindedness. Alfred Fairhurst, for example, argued, "I believe that the teaching of evolution is mostly dogmatic, and that the result of teaching it is a new crop of dogmatists. I am aware that there are those who hold that the subject of evolution greatly expands the mind. I think that, as taught, it warps the mind and closes it against much truth."[20] Baptist editor Curtis Lee Laws complained in the aftermath of the Scopes trial that the popular stereotype had taken over fundamentalism. "When will sensible people cease to revile and seek to understand fundamentalists?" Lee asked. He admitted that "there are among fundamentalists extravagant men who have made of themselves a public nuisance," but argued that the essence of fundamentalism ought to include tolerance and measured intellectual argument.[21] Had Bryan survived, could the weight of such arguments, added to Bryan's prestige and popularity, have managed to support wider definitional boundaries for fundamentalism? Could Bryan's progressive attitudes have limited the influence of the "extravagant men" such as Edward Young Clarke and J. Frank Norris who had embraced and successfully promoted the new image of fundamentalism?

Such questions, of course, can never be answered. As it happened, those fundamentalists who valued intellectual engagement with mainstream society found themselves squirming uncomfortably outside the boundaries of the new image of fundamentalism. Even worse, such ambivalent fundamentalists also found themselves closed out of mainstream progressive thought as well. Just as Dudley Field Malone had successfully articulated the defense at the Scopes trial as being the side of open-mindedness, so reformers such as John Dewey had successfully incorporated such ideas as secularism and material evolution within the boundaries of progressivism.[22] Those first fundamentalists who had become disenchanted with the new connotations of fundamentalism could not easily side with Dewey's vision of progressive educational reform. They could not overcome the consensus view of progressive identity as the domain of liberal and modernist Protestantism.[23] Nor could they continue to embrace fundamentalism as it shrunk to include only traditionalist, reactionary activism. They could not abandon their respect for the intellectual traditions of American evangelicalism, nor could they agree to the new mainstream intellectual presuppositions of materialism and theological liberalism.

This left these first fundamentalists in a difficult position. Not least among their difficulties, most fundamentalists who found themselves stranded outside the borders of both fundamentalism and liberalism could not or did not articulate their conflicted position. They found themselves contending with a confusing situation about which they could form no coherent analysis. A few fundamentalist educators, however, struggled successfully to make sense of the changing nature of fundamentalism. In Dallas, school founder Lewis Sperry Chafer and his colleagues argued explicitly about the changing nature of fundamentalism. Princeton theologian J. Gresham Machen also applied his academic acumen to a cogent analysis of the complex shifting boundaries. At Moody Bible Institute in Chicago, Gray worked along similar lines to diagnose the troubling changes in fundamentalism.

Chafer began his career as an itinerant Bible scholar decades before the Scopes trial. By 1906, he had dedicated his career to promulgating a dispensational premillennial understanding of the Bible. Like other conservative evangelicals, Chafer worried about the state of America's seminaries. By 1921, Chafer had decided to found a new fundamentalist seminary. His plan attracted influential supporters, such as theologian W. H. Griffith Thomas and Presbyterian missionary leader Alexander B. Winchester. By the next year, the three founders had agreed on a plan: to build a new type of seminary in Dallas, one that would "teach the fundamental doctrines of the unmutilated Word of God." Unlike Bible institutes, this seminary would insist on academic rigor. Unlike many contemporary seminaries, it would avoid divisive internal fundamentalist-modernist controversies by aligning itself adamantly with premillennial dispensationalist theology.[24]

Chafer and his close associates worried constantly about their relationship with the fundamentalist movement. In particular, they critiqued the meanings given to the term by leading fundamentalist voices such as Norris and Riley. Chafer and other Dallas founders attacked Norris in 1922, protesting against Norris's unwarranted condemnation of the recently published International Sunday School lessons.[25] Most troubling to Chafer and the other Dallas founders was the tone of fundamentalist rhetoric. Winchester frowned upon Norris's "angry polemics."[26] Many Dallas founders also condemned Norris's eager participation in the 1923 "trial" of three Methodist colleges in Texas. They presciently noted that such tactics would encourage an association of fundamentalism with medieval inquisitions. Chafer worried that such fundamentalist activism would smear the academic and intellectual reputation of his new school. As he wrote to prophecy writer and confidant Arno Gaebelein, "Because of the accurate meaning of that name [fundamentalism], we are coupled with all they do in the public mind."[27] Gaebelein shared similar fears. "The other side," he correctly predicted, "is going to use Mr. Norris and his reputation against the Fundamentalist Movement."[28] Chafer also protested against the theological inconsistencies of both Riley and Norris. Chafer charged that Norris was entirely too friendly with the Pentecostal movement, even publishing articles in the *Pentecostal Evangel*.[29] Riley, according to Chafer, had strayed away from the doctrine of Christ's "impeccability." That is, Riley believed that Christ was capable of sin, instead of essentially divine and thus incapable of human sin.[30]

In spite of these disputes with Riley and Norris, the leaders of the Dallas seminary often identified themselves as part of the fundamentalist movement. For instance, in his attack on Riley's theology, Chafer accused Riley of being someone who only "poses as a fundamentalist."[31] Chafer argued that his own theological steadfastness made him a more consistent representative of true fundamentalism. Elsewhere, Chafer dismissed Riley and Norris as merely "so-called Fundamentalists."[32] Chafer hoped to vindicate his own definition of fundamentalism as coequal with dispensational premillennial theology. In other contexts, Chafer acknowledged that his new school ought to be considered fundamentalist if the term were understood in an "accurate" sense.[33]

At times, these conflicts over the proper and exact use of the fundamentalist label led Chafer and his colleagues into some burdensome prose. As Chafer explained to Oliver Buswell of Wheaton College, the Dallas seminary leaders "stand for all of the fundamentals of the Word of God, [but] we are not identified with the Fundamentalist Movement as such."[34] Charles G. Trumbull, editor of the *Sunday School Times* and a close confidant of Chafer,[35] once described Chafer and other like-minded conservative evangelicals as "those who stand for the fundamentals of the faith (though not necessarily calling themselves technically 'Fundamentalist' or members of the Fundamentals Association)."[36] In an attempt to avoid describing themselves as simply "fundamentalists," Chafer and the other leaders of the Evangelical Theological College often resorted to practically unmanageable alternative terms. In 1925, for instance, they described themselves as "the company of those who demand authoritative, accurate preaching of the Word of God."[37] Elsewhere, the founders of the Dallas seminary identified themselves as "men who meet the conditions for spiritual power . . . for the defense of the faith delivered once for all time unto the saints."[38] Clearly, these labels were cumbersome, and Chafer often used simpler terms to identify himself, including simply "fundamentalist,"[39] but also "orthodox"[40] and "evangelical."[41]

Although the leaders in Dallas occasionally used the term to describe themselves, they shared a strong distaste for the growing anti-intellectual connotations of the fundamentalist label. They had founded their seminary, after all, to be a strong intellectual center for dispensational theology. In the aftermath of bruising battles over schools and school policy, the founders of the Dallas seminary, like fundamentalists nationwide, wrestled with the new connotations attached to fundamentalism. Rollin T. Chafer, Lewis Sperry's brother and the editor of the seminary's bulletin, complained in 1928 that secular "modern materialists" often blasted fundamentalist beliefs with "ridicule," but he predicted optimistically that fundamentalist beliefs would soon be defended by a new "Spirit-directed intellectual giant."[42] Rollin Chafer articulated some of the distress of intellectual fundamentalists, smarting from the barbs of secular critics such as H. L. Mencken. In Rollin Chafer's opinion, such secular pundits as Mencken and other "religious liberalists" had been occupied in "the debunking of the Christian faith," but such attacks could not withstand the intellectual power of traditional Christian apologetics.[43] By the end of the 1920s, Rollin Chafer bitterly noted that a "pitiable superiority complex" had seized secular and liberal intellectual culture. Instead of respecting the intellectual tradition of fundamentalists such as those at the Dallas seminary, these "modern scholastics" had reverted, in Chafer's opinion, to an ill-considered prejudice against all believers in "Christian supernaturalism." As did fundamentalists across the nation, Chafer and the other leaders of the Dallas seminary resented the sharply restricted boundaries associated with fundamentalism in the aftermath of the Scopes trial. However, due to the ways those new boundaries had been embraced by both liberal Protestants and by several leading fundamentalists, Chafer could find no way to expand those shrinking definitional boundaries.[44]

Due to their distaste for the narrowing definition of fundamentalism, however, the Chafer brothers and the other Dallas leaders had long been intellectually prepared to disavow the fundamentalist label if necessary. In a letter to leading conservative Presbyterian theologian Robert Dick Wilson, Lewis Sperry Chafer described his new school in revealing terms. The seminary, Chafer promised, was "quite independent of the Fundamentalist Movement . . . and the aim is to make it of the very highest class in every particular." On this occasion, Chafer negotiated the complicated boundaries of fundamentalism by identifying Riley's World's Christian Fundamentals Association (WCFA) as a capital-F "Fundamentalist" group.[45] Other Dallas founders similarly identified their disgust not with fundamentalism as it should be properly defined, but with "Riley's Fundamentalist movement."[46] In the end, however, Chafer recognized that this distinction no longer mattered. By the end of the 1920s, the fundamentalism of leaders such as Riley and Norris had become the commonly accepted definition of the fundamentalist movement as a whole. By 1930, Chafer admitted, "I have not been in sympathy with the movement from its beginning largely because of the fact that it has been negative in its testimony." [47]

Unable to articulate a fundamentalist self-understanding separate from the new image of the fundamentalist movement, Chafer eventually resigned himself to an uncomfortable ambiguity and inexactness in terms of his relationship to the fundamentalist movement. He had struggled throughout the decade to dispute the boundaries of fundamentalism as asserted by Norris and Riley. By the decade's end, however, he recognized that such boundaries had become the consensus understanding. He stipulated that he had not changed his beliefs—even those conservative tenets that had originally attracted him to the fundamentalist movement—but that he no longer could be part of a movement that had utterly lost its mainstream intellectual respectability. In short, Chafer believed that he had remained true to the original meanings of fundamentalism, but he recognized that the consensus understanding of that term had changed. He could no longer define his own position within the newly constricted boundaries of fundamentalism. Nevertheless, he could and did continue his educational activism, building his school into one of the institutional homes of the fundamentalist movement.

Conservative Presbyterian theologian J. Gresham Machen struggled similarly with his relationship to the fundamentalist movement throughout the decade. At times, Machen implicitly accepted a fundamentalist label. By the end of the decade, however, he articulated a nuanced self-definition that described his position in relationship to fundamentalism. Since the publication of Machen's *Christianity and Liberalism* in 1923, many fundamentalists and liberals had considered Machen the leading intellectual of the fundamentalist movement.[48] A careful reading of this work, however, would have given some fundamentalists pause. Machen located the educational problem of American schools not in their growing acceptance of evolution and secularism, but rather in their increasing materialism and intellectual sterility. In language more similar to H. L. Mencken's than to that of most other fundamentalists, Machen attacked the majoritarian rule of schools. The problem with schools, Machen concluded, mirrored that of American society at large. He damned schools as restricting freedom of thought. In the past, Machen argued, such restriction on free learning had been carried out by "the Inquisition, but the modern method is far more effective." Machen worried that academically inferior schools "will rapidly make of America one huge 'Main Street,' where spiritual adventure will be discouraged and democracy will be regarded as consisting in the reduction of all mankind to the proportions of the narrowest and least gifted of the citizens."[49]

Nevertheless, Machen allowed himself to represent the intellectual cutting edge of the fundamentalist movement. He often described himself as articulating the ideas of the movement "in a broad sense."[50] As the Scopes trial approached, Machen made his struggle to maintain the mainstream intellectual status of the fundamentalist movement explicit. He argued that the label had become "distasteful" to him, since it was so often used inappropriately to imply that conservative evangelicals had joined "some strange new sect." To the contrary, Machen argued, his vision of fundamentalism implied nothing more radical than "simply . . . maintaining the historic Christian faith and . . . moving in the great central current of Christian life." For Machen and many similarly ambivalent conservatives, continued identification with fundamentalism threatened exile from mainstream religion and culture.[51]

After the inexorable pressure on the popular image of the fundamentalist movement in the wake of the Scopes trial, Machen continued to attempt to rescue the term while simultaneously asserting the term's inadequacy. In 1927, for instance, when pressed for a self-definition, Machen offered the following heavily qualified response: "I never call myself a 'Fundamentalist.' . . . if the disjunction is between 'Fundamentalism' and 'Modernism,' then I am willing to call myself a Fundamentalist of the most pronounced type. But, after all, what I prefer to call myself is not a Fundamentalist but a Calvinist."[52] Machen had always rejected both the popular image of fundamentalism as an anti-intellectual movement and the theological image of fundamentalism as a movement restricted to belief in dispensational premillennialism. But if fundamentalism meant a coalition of activist evangelicals combating theological modernism, then Machen considered himself a leading member. In the aftermath of the Scopes trial, Machen bitterly agreed that "the noisy pretentions [sic] to superior scholarship that many of the liberals make" had managed to score important successes. In Machen's view, such liberal attacks had pushed some conservative evangelicals to adopt an anti-intellectual posture unwisely.[53]

In the late 1920s, Machen gamely fought to reassert his preferred terms of "conservative" and "evangelical" as correct labels for his activism. He hoped to retain with both terms the tradition of elite scholarly accomplishment inherited from the likes of Puritan scholar Jonathan Edwards. However, he found that both his enemies and supporters ignored his nuanced self-definition.

Part of Machen's difficulties of self-definition resulted from his continuing need to maintain a polite identification with the new image of the fundamentalist movement. Unlike other leading fundamentalists, Machen needed the support of self-identified Presbyterian fundamentalists due to the intensification of Machen's educational activism. By the end of the 1920s, Machen decided that he could no longer remain affiliated with Princeton Theological Seminary due to the increasing influence of theological modernism at that institution. He and other conservatives split off and founded Westminster Seminary. Machen hoped that the new school would be a "truly evangelical seminary," free from the influence of theological modernists.[54] Although Machen did not relish his association with the new image of fundamentalism, he recognized that he needed support from Presbyterian conservatives to make his new seminary a success. Machen recognized that many conservative Presbyterian activists continued to consider him the leader of Presbyterian fundamentalists. During the late 1920s, as he worked to establish the new seminary, he maintained an awkward balancing act by accepting the support of self-styled fundamentalists without explicitly identifying himself as such. He pointedly avoided using the fundamentalist label, even when his supporters continued to identify themselves proudly as fundamentalists. For instance, when one ardent supporter pressured him to lead the "Fundamentalist ticket" at the 1928 Presbyterian General Assembly,[55] Machen thanked him for his enthusiasm in the "conservative" or "evangelical" cause."[56] However, both his friends and enemies among conservative Presbyterians identified Machen with fundamentalism. One Presbyterian opponent asserted that Machen's personal attacks had done a great deal to "discredit Fundamentalism" among leading Presbyterian intellectuals.[57]

Like Chafer, Machen wrestled with the new meanings of fundamentalism. He refused to accept the label's connotations of anti-intellectualism and mob rule. Yet he continued to identify with the conservative evangelical Presbyterianism that had been one of the key players in the denomination's fundamentalist-modernist controversy. He waged his militantly conservative educational battles throughout the 1920s with the ardent support of self-styled fundamentalists, but he hoped to do so without personally accepting the fundamentalist label.

James Gray of Chicago's Moody Bible Institute articulated a similarly complicated relationship to the fundamentalist movement in the years following the Scopes trial. Gray had embraced fundamentalism in the early 1920s as a successful revival of conservative evangelical Bible-based culture. After the Scopes trial, Gray found himself in a similar position to Machen and Chafer. As an urban Northern academic, Gray felt extremely uncomfortable within the new image of fundamentalism. He chafed at accusations of bigotry and ignorance, and struggled to maintain a wider definition for fundamentalism. When he could not, he quietly allowed himself to slip outside the ranks of self-identified fundamentalists.

When Pittsburgh steel executive Thomas Gillespie took the stage at the 1927 Bible conference to protest his belief that fundamentalists were indeed the real progressives, Gray enthusiastically agreed. To his readers in the *Moody Bible Institute Monthly*, Gray called Gillespie's remarks "so apt and striking, and presented so frank a challenge to the deriders of the conservators of the Christian faith, that we are happy to be able to give them to our readers verbatim." Gillespie, in Gray's opinion, managed to articulate a vision of fundamentalism beyond the popular new stereotype. Gray agreed that the "deriders" of fundamentalism had seized the upper hand in the definitional debate over fundamentalism. As did Gillespie, Gray desperately hoped to maintain wider boundaries for the term.[58]

Gray also encouraged his students to resist attacks from liberals, rather than to accept those accusations as badges of honor, as some fundamentalists had done. "Let them not shame you with the taunt of narrowmindedness," he advised. Rather, work to build an evangelical identity beyond the one imposed by liberals and embraced by fundamentalists such as Norris.[59] Gray repeatedly warned his readers that liberal critics often misrepresented fundamentalism.[60] In the aftermath of the Scopes trial, he warned his readers to "be on their guard against newspaper arguments opposing the Christian faith and not to be carried away by them."[61]

In the end, however, Gray quietly accepted that he could not alter the new consensus about the meanings of fundamentalism. He slowly repositioned himself as an ally outside the boundaries of fundamentalism. He used his influence to make sure self-identified fundamentalists were welcome at the Moody Bible Institute (MBI), but that the institution itself claimed a wider evangelical label. For instance, he continued throughout the 1920s to allow self-identified fundamentalists to defend the movement in the pages of MBI publications.[62] But Gray himself clearly preferred other labels. Whereas fundamentalists such as Norris and Riley began to use the label fundamentalist much more aggressively after the Scopes trial, Gray slowly reverted to using other terms. For example, in 1926, Gray's *Moody Bible Institute Monthly* supported the leaders of the Baptist Bible Union as "fundamentalist" activists.[63] The next year, the magazine described the same men only as "speakers of international reputation."[64] Gray himself switched from using the term fundamentalist to such terms as "defenders of the Christian faith,"[65] the "orthodox,"[66] and even "conservative scholars."[67] At times, this terminology became noticeably cumbersome. For instance, when giving his support to the self-identified fundamentalist leaders of the new

Des Moines University in 1927, Gray would only describe them as "men who know and believe in the God of the Bible and who will not compromise with his enemies."[68] Most remarkably, he publicly acknowledged his willingness to abandon the fundamentalist label in 1927. "'Fundamentalism' as a slogan may go," he wrote, "but that which gives it reason can never go."[69] Although Gray did not want to position himself as an opponent of fundamentalism, he had decided that the movement as generally understood after 1925 no longer had room for his kind of scholarly Bible-centered evangelicalism.

Unfortunately for Gray, Machen, Chafer, and the many grassroots fundamentalist activists who agreed with their careful logic, there was no credible alternative label around which they might rally. These early fundamentalists found themselves outside the newly restricted boundaries of fundamentalism. However, they also disavowed the beliefs of fundamentalism's liberal opponents. Even if they considered themselves progressives, they could not stomach the imputation of secularism that such a label connoted in the 1920s. The shrinking boundaries of fundamentalism left these fundamentalists without a compelling understanding of their proper role in wider American politics and culture. Most of them likely agreed with Machen's careful self-definition. They continued to consider themselves fundamentalists, but only in the sense that they continued to loath the tenets of liberalism and theological modernism. In spite of such continuing strong feelings, they shied away from association with any public campaign identified with fundamentalism.

Nevertheless, they did not necessarily cease their activism entirely. Many of the first fundamentalists continued to fight for reforms that had not been wholly identified with fundamentalism. For instance, many leaders at the Moody Bible Institute and elsewhere continued their campaign to introduce Bibles and religious literature in public schools. Lewis Sperry Chafer and J. Gresham Machen continued to fight for their vision of Christian education. Others continued their missionary efforts unabated. Such campaigns had not become as thoroughly equated with fundamentalism as the antievolution campaign. In the light of the rapidly restricting boundaries of the fundamentalist movement, such campaigns appealed to those early fundamentalists who had struggled, like Thomas Gillespie, in the wake of the Scopes trial to understand what it meant to be a fundamentalist who considered himself "progressive."

Conclusion

Contemporary observers disagreed about the net result of the fundamentalist school campaigns of the 1920s. Maynard Shipley, self-styled president of the Science League of America, concluded glumly in 1930 that he and other antifundamentalist activists had lost their war. "Nothing can be taught," Shipley warned, "in 70 per cent of the secular schools of this Republic today not sanctioned by the hosts of Fundamentalism."[1] Other foes of fundamentalism reached the opposite conclusion. Liberal Presbyterian intellectual William Adams Brown surmised with relief in late 1926 that the fundamentalist movement was "on the ebb" and that its energetic activism would not have "the serious consequences once anticipated."[2]

This confusion about the net result of fundamentalist school activism resulted in large part from the complex nature of the fundamentalist movement and the ambitious scope of its school campaigns. The debates over whether fundamentalists won or lost their fights in the 1920s are similar to the old story of the blind wise men and the elephant. In the story, each blind man examines only one part of the elephant and comes to a different conclusion about what an elephant must be. One blind sage feels the elephant's trunk and concludes that an elephant resembles a thick snake. A scholar who examines the elephant's leg disagrees; he argues that the elephant is more similar to a tree stump. On the other hand, the blind man who is most interested in the elephant's tusk asserts that an elephant is not an animal at all but a kind of smooth stick. Like those scholars, observers reached different conclusions about the overall success or failure of fundamentalism based on their focus on different parts of the many-headed fundamentalist movement of the 1920s.

On one hand, the movement scored remarkable successes. Fundamentalists helped pass several state laws banning the teaching of evolution and securing preferential status for traditional Protestant theology in public schools. Those observers, like Maynard Shipley, who focused on the fight against evolution could justifiably conclude that the fundamentalists had won. Although fundamentalist campaigns for state laws often failed, their influence on local policy and textbook publishing meant that most American schoolchildren received relatively little explicit education in evolutionary theory until the 1960s.[3]

Even beyond the fight against evolution, fundamentalists achieved remarkable success during their 1920s controversies. Fundamentalists and their allies legally mandated the reading of the Bible in the public schools of several states. More important, in the case of America's diffuse educational system of locally run schools, fundamentalists pushed successfully for local laws promoting their vision of religiously, culturally, and politically appropriate public education. Moreover, fundamentalists in the 1920s founded or gained the allegiance of a powerful network of independent colleges, seminaries, elementary and secondary schools. Few educational movements in American history can make such claims.

On the other hand, the school campaigns tore the fundamentalist movement apart. Even their successes forced fundamentalists to come to terms with their new public image as a group of what Clarence Darrow termed "bigots and ignoramuses."[4] The publicity from ambitious educational campaigns and especially from the Scopes trial branded the fundamentalist movement with an image as a group of ignorant Southern reactionaries. Some fundamentalists, as we have seen, helped cement the new stereotype by defiantly embracing a fundamentalist image that touted its populist defense of traditional Southern values. In light of this transformation in the meanings of fundamentalism, those activists, like William Adams Brown, who focused on fundamentalism's slide from mainstream intellectual respectability could also claim convincingly that the movement had lost its 1920s campaigns.

As with the story of the blind men and the elephant, debates over whether fundamentalists won or lost their battles address only discrete parts of the movement's experience in the 1920s. The successes and failures of the fundamentalist school campaigns of the 1920s were bound together inextricably. This book's focus on those school campaigns in all their complexity has sought to demonstrate the relationship between these varied campaigns. Moreover, a focus on fundamentalist educational activism as a whole also clarifies the ways that fundamentalist school battles led to dramatic transformations of both the fundamentalist movement and the American educational system itself.

At the start of their school campaigns, most fundamentalists regarded themselves as the voice of mainstream American cultural opinion. This foundational belief in the popularity and mainstream legitimacy of their school campaigns buckled under the weight of unexpected attacks on fundamentalists' role as intellectual and religious leaders. Fundamentalist confidence did not shrink from the vigorous opposition they faced—that was generally expected—but from the way their opponents often seized the moral and intellectual high ground in public debate. With widespread popular success, liberals and secularists depicted these school campaigns as a fight between the modern world and a world locked in the past. Even when fundamentalist school policy triumphed, these accusations lingered, bolstered by the defensive embrace of the new image by many leading fundamentalists. As some observers had portrayed locals at the Scopes trial as primitive zealots, so the enemies of fundamentalism portrayed the entire movement as a sort of menacing, outdated eruption of ignorance.

Fundamentalists were forced to choose between a defensive, inward-looking embrace of the new meanings of their movement and a quiet and ambivalent abandonment of the fundamentalist label. This unwelcome decision between two imperfect choices changed fundamentalism permanently. Beginning in the early 1920s as a relatively broad coalition of Protestants, fundamentalism became by the end of the decade a narrower band of defiant outsiders determined to remain true to the fundamentalist label despite the resulting exile from mainstream intellectual respectability.

Further, the transformation in the fundamentalist movement both caused and resulted from similarly dramatic changes in American education as a whole. Fundamentalists discovered through their 1920s school campaigns that they could not unilaterally dictate educational policy and culture. They found that traditional Protestant belief no longer represented the accepted—if unofficial—curriculum of American schools. Fundamentalist activists realized to their surprise that their idea of appropriate education met with determined and sustained resistance from much of the new educational establishment. This discovery introduced a potent and long-lasting new interest group into American education. Instead of a movement confident in its own place in the educational establishment, fundamentalism—and eventually the wider conservative evangelical Protestantism from which fundamentalism developed—became a movement fighting for its place at the policy table. Though fundamentalists and conservative evangelicals in some regions may have lost their traditional ability to dictate public school policy, they remained a potent political force for educational policy. Fundamentalists fought for control of public schools with a new understanding of their role as aggrieved outsiders. Or, if convinced that they could not keep public schools safe for their own children, fundamentalists retreated to a new independent school system of their own. Schools such as Bob Jones University and Dallas Theological Seminary became vital institutional centers for independent fundamentalist schools and churches throughout the twentieth century. The allegiance of existing schools such as Moody Bible Institute and Wheaton College strengthened this new educational system.

By introducing both a new parallel school system and a powerful new interest group in public education, the 1920s fundamentalist school campaigns effected a change in American education as a whole that lasted throughout the twentieth and into the twenty-first centuries.

In short, the school controversies of the 1920s introduced new lines of conflict into perennial culture wars over education. As opposed to nineteenth-century struggles over schools that pitted Catholic activists against a staunchly Protestant educational establishment, twentieth-century battles—which began in earnest with the ambitious and varied fundamentalist school campaigns of the 1920s—matched liberals of many religious backgrounds against their conservative coreligionists. The public schools no longer represented a meeting place acceptable to most Protestants. Instead, fundamentalists defined themselves, and were defined by their opponents, as an active new interest group outside of the educational mainstream. Their victories gave them confidence that their ideas represented the educational philosophy of a large segment of tradition-minded Americans. But fundamentalists' many losses taught them that their vision of proper education no longer dominated America's educational establishment. No longer supremely confident in claiming their rights as part of a unified Protestant educational establishment, fundamentalists took up lasting positions as usurped outsiders. As outsiders, fundamentalists followed the path of other minority groups in American education. Fundamentalists, like other minority groups, created their own network of schools. Just as with other groups, they proudly claimed some regions of continued majority rule. They also stridently demanded changes in the curricula and culture of those public schools that they could no longer dominate. Unlike other minority groups, however, fundamentalists pointedly remembered—sometimes with significant wistful embellishment—a past in which their beliefs had dominated America's educational mainstream.

Examined as a whole, the fundamentalist school campaigns of the 1920s led to both these vital transformations in American life. Fundamentalism itself changed from a relatively wide coalition of Bible-centered Protestants to a much narrower subculture of defiant traditionalism. In addition, the American educational system changed to transform fundamentalists into a powerful new interest group determined and able to contend bitterly in the contested twentieth-century field of pluralistic public education.

EPILOGUE

Into the Future

In December 1981, a stream of reporters, theologians, scientists and gawkers converged on the federal courthouse in Little Rock, Arkansas. As the Scopes trial had for their grandparents' generation, the trial of *McLean v. Arkansas* promised a showdown between the rationality of science and the entrenched beliefs of fundamentalist religion. The issue was an Arkansas law mandating equal time for the teaching of scientific creationism and evolution in public schools. Not surprisingly, many observers called the trial "Scopes II."[1] The *New York Times* even reprinted 1925 commentaries by both H. L. Mencken and Clarence Darrow.[2]

As seen from one perspective, the equation of the two trials did not make much sense. It is true that they both focused on the issue of the teaching of evolution in public schools, but the positions of the two sides had largely reversed themselves. In 1925, fundamentalists battled to maintain Protestant control of schooling; in 1981, they fought to be included. In 1925, they hoped to ban any doctrine, especially including evolution that challenged students' Protestant beliefs. Noting that their position had "essentially turned around 180 degrees," leading scientific creationists by the end of the twentieth century fought not for dominance, but merely to be heard. As did other creationists, pundit Duane Gish pointed out that "Clarence Darrow thundered that it was bigotry to teach only one theory of origins."[3] (Gish was mistaken in his attribution; it was attorney Dudley Field Malone who made that claim at the Scopes trial, not Darrow.)[4] Gish now intimated that it was bigotry to exclude creationism from the curriculum in public schools. Fundamentalist educational activists had exchanged William Jennings Bryan's majoritarian arguments for pleas to respect for the minority rights of fundamentalist Protestants. Considering this trial a repeat of the 1925 experience in Dayton, Tennessee, ignored this important transformation.

In another sense, however, calling the Arkansas trial Scopes II was entirely apt. The trial represented the continuing culture wars over the role of evangelical

Protestant religion in American education. Events surrounding those conflicts could not have unfolded as they did without the educational controversies in the 1920s. The changes in both fundamentalism and schooling wreaked by those 1920s struggles determined the positions and the issues that would fuel the educational culture wars of the later twentieth century. For instance, the new network of fundamentalist schools opened and consolidated in the 1920s led to a powerful and durable fundamentalist subculture. Yet fundamentalists never abandoned activism in public schools. They continued to exert pressure on public schools throughout the century. By the 1960s, many key 1920s fundamentalist victories, such as the inclusion of Bibles in public schools and the prohibition of evolution from those schools, came under new attack. The educational struggles of the 1960s and later decades resulted largely from evangelical activism to preserve those hard-fought privileges. Even fundamentalist defeats in the 1920s transformed the movement into a new type of educational movement, one that eventually embraced the "rights" rhetoric of Dudley Field Malone and Clarence Darrow at the Scopes trial. In this sense, the Arkansas trial in 1981 certainly fits the description of sequel to the 1925 trial. As one flare-up in the culture wars of the later twentieth century, the Arkansas trial continued the cultural conflict that had been inaugurated so vividly during the school controversies of the 1920s.

One of the reasons for that enduring conflict was the continuing division between Americans of differing religious and scientific beliefs. At the end of the twentieth century, according to Gallup poll results, just under half of American adults agreed that "God created human beings pretty much in their present form at one time within the last 10,000 years or so." That percentage included roughly one quarter of all Americans with college degrees.[5] This durable belief was not due to lack of education, as some critics might have charged, but by the development of America's educational system since the 1920s. Due to the controversies of the 1920s, American public schools remained largely free of evolutionary theory until the 1960s, as described below. Even more important, fundamentalists in the 1920s began to build a powerful network of independent schools. By the late twentieth century, many fundamentalists and evangelicals had received their education exclusively at schools and colleges dedicated to an evangelical, Bible-based theology and science. Thanks to those schools, one could be both well educated and thoroughly steeped in biblical creationism.

Begun during the controversies of the 1920s with new colleges such as Bryan College and Bob Jones College, as well as the new explicitly fundamentalist allegiance of existing schools such as Wheaton College, generations of fundamentalist educators have expanded the independent evangelical school network to become one of the largest private school systems in the United States. According to one journalist's estimate, three new fundamentalist or evangelical schools were opened every day in the United States throughout the late 1970s and early 1980s.[6] By 1975, approximately 400,000 students enrolled in evangelical and fundamentalist K-12 schools.[7] By 2002, that number had more than doubled with over 800,000 students enrolled. Moreover, independent fundamentalist and evangelical schools had become, by 2002, the largest single category of private schools with 5,527 schools. By way of comparison, there were 4,347 parochial Catholic schools in 2002, plus 2,933 diocesan Catholic schools, and 2,939 "regular" nonsectarian private schools.[8]

Although journalists and other observers noted the explosion in numbers of these "Christian day schools" only in the 1970s, the schools would not have sprung into existence were it not for decades of work by fundamentalist educational activists. An example of these direct connections has been the experience of Bob Jones University (BJU). Begun in response to the controversies of the 1920s, as described in Chapter 7, BJU has played a decisive role in shaping the century's cultural conflicts over schools. Much of the institutional foundation of the burst of Christian day schools in the 1970s resulted directly from the work of activists associated with BJU.[9]

For instance, in 1954, two BJU alumni founded their own Christian college. Arlin and Beka Horton opened Pensacola Christian College in order to train a new generation of fundamentalist classroom teachers. Eventually, their educational approach proved so successful they began a line of educational publishing. Their books and curricular materials remain leading choices for fundamentalist schools across the nation. Without such curricular materials to draw upon, the new explosion of evangelical schools could never have occurred.[10]

Similarly, in 1970 another BJU alumnus, Don Howard, began a new fundamentalist school publishing enterprise. Howard's Accelerated Christian Education curriculum (ACE) became a fast favorite for the new schools of the 1970s. Howard's program offered low-cost, theologically and culturally safe curricula for evangelical and fundamentalist schools.[11] Although derided as "skeletal . . . distorted . . . almost paranoid . . . very limited and sometimes inaccurate" by nonfundamentalist critics, the ACE curriculum fueled a good deal of the spectacular growth of the Christian day school movement.[12] The fact that secular scholars found fundamentalist school curricula to be inadequate should not come as a surprise. As ACE leader Ronald E. Johnson noted, he did not want to create a school curriculum with similar ideas to those taught in secular schools. Rather, Johnson, Howard, the Hortons, and other fundamentalist educators planned to create an entirely separate system, separate even from the basic cultural assumptions of mainstream American culture. Their successful campaigns later in the twentieth century relied directly on the ambitious efforts of 1920s activists such as school founder Bob Jones Sr.[13]

But even with their own thriving network of independent schools, fundamentalists and other conservative evangelicals did not by any means abandon the fight to control public schooling after the controversies of the 1920s. During the 1930s, 1940s, and 1950s, they did not attract the same media attention that they had during the controversies of the 1920s, or that they would in the struggles of the 1960s, largely because during those middle decades fundamentalists largely succeeded in keeping public schools friendly to traditional Protestant belief. Doctrines perceived as pernicious, such as evolution, largely remained excluded. Bibles, prayer, and other expressly Protestant expressions of belief remained embedded.

Until 1960, due mostly to self-censorship by textbook publishers, only a minority of the students in American public schools were exposed to the ideas of human evolution.[14] Similarly, Bibles and prayer remained a vital part of public school education in much of the country. A 1949 survey found that Bible-reading formed a part of public school education in at least thirty-five states.[15] Of the twelve states that legally required Bible reading, eleven had passed such laws between 1913 and 1930. These laws were not relics of a distant Puritan past, but rather part of the way fundamentalists and their allies made conservative Protestant beliefs part of the explicit, legal nature of public schooling during the 1920s. On the cusp of Supreme Court decisions in 1962 and 1963 banning prayer and Bible-reading from public schools, such practices remained common. In the South, almost 80 percent of elementary teachers reported Bible-reading as a regular part of their school days in 1960. Although those numbers were much lower in the West and Midwest, at 11 percent and 18 percent, respectively, they were almost as high in the East, with 67 percent of surveyed teachers reading from the Bible during their public school day.[16]

These traditional practices came under renewed legal pressure during the 1960s. One of the most influential changes was the introduction of evolution into many of America's schools. Beginning in 1960, a few years after the Soviet Union's *Sputnik* satellite spurred the federal government to improve science education, textbooks published by the federally funded Biological Sciences Curriculum Study (BSCS) made inroads into many public school districts. These textbooks featured evolution as one of their key themes. By the end of the 1960s, nearly half of American high schools used BSCS materials to some extent. Equally important, other publishers rushed to update their treatment of evolution in order to compete with BSCS textbooks. As a result, many more public school students used textbooks that treated the subject of human evolution thoroughly and explicitly. Many evangelical parents reacted with alarm to these curricular changes.[17]

By 1960, a new generation of fundamentalist educational activists such as Henry Morris had been working for years to promote the teaching of creationism in public schools. Based in part on popular conservative outrage at the new BSCS science curriculum, Morris and others began a drive to allow for a "two-model" approach in public schools. In this approach, both evolution and a Genesis-friendly creation science would be taught side by side. Throughout the 1970s and 1980s, Morris's strategy enjoyed great success. The state legislatures of Arkansas and Louisiana, along with many local school boards, adopted the two-model approach.[18] Controversy over the Arkansas law led to the 1981 Scopes II trial in Little Rock. In 1987, the Supreme Court struck down both the Arkansas and the Louisiana laws, but the issue did not go away. Even the controversial 2001 federal school law, the No Child Left Behind Act, included a nonbinding congressional conference report that the "full range of scientific views" should be taught whenever "controversial topics" are discussed. In other words, whenever public schools taught evolution, the federal government encouraged them to teach creationism as well.[19]

Just as the burst of new evangelical schools did, these later disputes over religion, evolution, and schooling had their roots in the school battles of the 1920s. The generation of fundamentalist scientists fighting for "creation science" in the 1970s and 1980s had been nurtured in the independent fundamentalist scientific establishment founded in the 1920s. When early fundamentalist scientists such as George McCready Price, Arthur I. Brown, and Harry Rimmer grew disheartened at their inability to convince obdurate mainstream scientists of the superiority of the fundamentalist approach, as described in Chapters 5 and 9, they founded new scientific institutions free from mainstream influence. As soon as a young Henry Morris, inspired by the work of Price and Rimmer, became passionate about fundamentalist science in the early 1940s, he was able to find moral support and an institutional home in Price's Deluge Geology Society. The work of 1920s-era fundamentalist scientists gave the succeeding generations both training in the tenets of their biblically guided science and a readymade institutional network separate from the presuppositions of mainstream science.[20]

These institutional connections demonstrate the direct and vital relationship between Henry Morris's activism in support of teaching scientific creationism in public schools and 1920s fundamentalist school activism. Similarly, the institutional supports that allowed a burst of new evangelical schools in the 1970s had their origins in the 1920s educational activism of school founders such as Bob Jones Sr. Just as important, when evangelicals and fundamentalists in the 1960s and 1970s responded to the changing legal status of prayer and Bible-reading in public schools, they relied directly on foundations laid in the controversies of the 1920s.

That renewed evangelical activism in favor of Bibles and prayer came in response to the Supreme Court's 1963 decision in *Abington School Dist. v. Schempp.*

In this case, Unitarian parents sued a school district for creating an atmosphere in which their son felt pressured to participate in the reading of ten verses from the King James Version of the Bible. After many appeals (the case was first brought in 1956), the Supreme Court agreed. Henceforth, no public school could continue the policy of mandatory reading of the Bible, even if individual exemptions were granted.[21]

Of course, Supreme Court decisions do not necessarily equate with local educational policy. As political scientist Kenneth Dolbeare has demonstrated, many school districts continued their traditional religious practices in spite of the *Schempp* decision.[22] Nevertheless, as did Americans of many religious backgrounds, evangelicals reacted to the decision with shock and outrage. A *Moody Monthly* poll of evangelical editors in early 1964 found that they considered the *Schempp* decision the most important event of 1963, outranking the year's civil rights activism and Birmingham's 16th Street Baptist Church bombing in importance to American culture and society.[23]

In spite of the Court's decision, the public school missionaries from Chicago's Moody Bible Institute (MBI) planned to continue their delivery of Bibles and evangelical literature to public schools. In the immediate aftermath of the decision, MBI activists correctly predicted the likely significance of the *Schempp* decision. "In general," they concluded, "states that have permitted or required Bible reading will probably continue the custom unless or until there is local protest. Then it will probably stop, not because of federal troops coming to haul away the Bibles from the schoolrooms, but because state departments of education will remove local accreditation." Nevertheless, the Chicago Bible missionaries promised to continue their distribution of Bibles to public schools. In fact, like many evangelical and fundamentalist school activists, they vowed to increase their program of energetic literature outreach to schoolteachers, principals, and superintendents, and to work to guarantee that every student heard the Bible during his or her school day. As did much of the evangelical and fundamentalist school activism of the culture wars of the late twentieth century, the Bible distribution did not represent a new program, but rather a continuation of a program that had begun during the school controversies of the 1920s.[24]

That did not mean that such activism continued unchanged from its roots in the 1920s. On the contrary, the 1920s experience of school activism often changed fundamentalists' strategies and tactics in decisive ways. As fundamentalists in the 1920s wrestled with their new stereotyped public image as rural, anti-intellectual hillbillies, they often adopted the tone and tactics of an aggrieved minority. As historian Joel Carpenter has argued, fundamentalists in the 1930s and 1940s often retreated from public controversy to build their own subcultural institutions and minority identity.[25] That sense of beleaguered minority activism emerged later in the century as evangelical and fundamental-ist activists fought again for control of public schools, especially over such issues as religious instruction, prayer, evolution, and sex education.[26]

The example of the Chicago book missionaries is illustrative of this transfor-mation. In the 1920s, as described in Chapter 8, the MBI's Bible-distribution program targeted public schools in isolated rural pockets of the country, espe-cially southern Appalachia. They hoped to deliver Bibles and the fundamental-ist message in such regions because they considered fundamentalism to be the belief of the overwhelming majority in such rural communities. After World War II, however, the Chicago Bible missionaries conceived of their mission in a new way. As did many evangelical educational activists, the MBI program began to compete for the souls of students in what they recognized to be pluralistic pub-lic schools across the country. Whereas in the 1920s the MBI activists targeted isolated "mountain" schools, after World War II they described their target as a nationwide body of "educators, alarmed over the morally destructive effect of the obscene literature which is being read by elementary school children."[27]

After World War II, the Chicago program abandoned its earlier desire to avoid controversy. Instead, the Chicago activists envisioned their evangelical message as but one voice public school students would hear during their school days. They hoped to reach students in city and suburban schools, painfully con-scious that many of those students no longer shared the cultural and theological background of the Moody Bible Institute evangelists.[28]

As did much of the educational activism of the late twentieth-century cul-ture wars, this transformation from an insistence on the majority rights of evan-gelical Protestants to a strategy of battling for evangelical inclusion in public schools had its roots in the bitter struggles of the 1920s. During that decade, fundamentalists fought to assert their cultural control over public schooling. As they learned to their dismay that they could no longer simply shame their oppo-nents into compliance, the broad coalition of conservative evangelicals that made up the early fundamentalist movement endured important transforma-tions. The meanings of fundamentalism itself changed to imply a much smaller subculture of intolerant, backward-looking conservatives. The struggles over the nature of education changed. Those changes did more than fill the head-lines of 1920s newspapers. They led to durable new realities in both education and evangelicalism. Those new definitions laid the foundations for the cultural struggles over God and school that lasted throughout the twentieth century.

Notes

Introduction

1. Edward Larson, *Summer for the Gods: The Scopes Trial and America's Continuing Debate over Science and Religion* (Cambridge, MA: Harvard University Press, 2001), 187–90.
2. Trial transcript from Jeffrey P. Moran, *The Scopes Trial: A Brief History with Documents* (Boston: Bedford/St. Martin's Press, 2002), 156.
3. Larson, *Summer for the Gods*, 239–46.
4. James Davison Hunter, *Culture Wars: The Struggle to Define America* (New York: Basic Books, 1991), 43 [italics in original].
5. See David B. Tyack, "Onward Christian Soldiers: Religion in the American Common School," in *History and Education: The Educational Uses of the Past*, ed. Paul Nash, 212–55 (New York: Random House, 1970); Donald E. Boles, *The Bible, Religion, and the Public Schools* (Ames, IA: Iowa State University Press, 1963); James C. Carper and Thomas C. Hunt, *The Dissenting Tradition in American Education* (New York: Peter Lang, 2007); Warren A. Nord, *Religion & American Education: Rethinking a National Dilemma* (Chapel Hill, NC: University of North Carolina Press, 1995).
6. Morris P. Fiorina, Samuel J. Abrams, and Jeremy C. Pope, *Culture War? The Myth of a Polarized America* (New York: Longman, 2004). See also Benjamin Justice, *The War that Wasn't: Religious Conflict and Compromise in the Common Schools of New York State, 1865–1900* (Albany, NY: SUNY Press, 2005); David Brooks, "The Middle Muscles In," *New York Times*, November 9, 2006, A33.
7. Ward W. Keesecker, *Legal Status of Bible Reading and Religious Instruction in Public Schools* (Washington, DC: United States Government Printing Office, 1930), 2.
8. W. C. Howells, "Donahey Defies Klan in Vetoing Bible Measure," *Cleveland Plain Dealer*, May 1, 1925, 1, 3.
9. W. S. Fleming, *God in Our Public Schools*, 3rd ed. (1942; repr., Pittsburgh, PA: National Reform Association, 1947), 143–44. The following states (with the year the law passed) required daily Bible reading: Massachusetts (1826), Pennsylvania (1913), Tennessee (1915), New Jersey (1916), Alabama (1919), Georgia (1921), Delaware (1923), Maine (1923), Kentucky (1924), Florida (1925), Idaho (1925), Arkansas (1930).
10. Mark Taylor Dalhouse, *An Island in the Lake of Fire: Bob Jones University, Fundamentalism, and the Separatist Movement* (Athens, GA: University of Georgia

Press, 1996), 152–53; Daniel L. Turner, *Standing without Apology: The History of Bob Jones University* (Greenville, SC: Bob Jones University Press, 1997), 15–21, 263–69. See also Adam Laats, "Inside Out: Christian Day Schools and the Transformation of Conservative Protestant Educational Activism, 1962–1990," in *Inequity in Education: A Historical Perspective*, ed. Debra Meyers and Burke Miller (Lexington, KY: Rowman and Littlefield Press, 2009), 183–209.

11. Timothy Weber, *Living in the Shadow of the Second Coming: American Premillennialism, 1875–1982* (Chicago, IL: University of Chicago Press, 1987), 238.

12. See *New York Times* coverage, including the following: "Paint W. J. Bryan as a 'Medievalist,'" March 2, 1922, 12; "Bryan Renews War on Darwin Theory," April 14, 1923, 6; "Anti-Darwin Campaigns Stir South and West," June 10, 1923, X2; "Assail Darwinian Theory," August 1, 1924, 26; "Science and Religion," April 5, 1925, E4; "Evolution Trial Raises Two Sharp Issues," May 31, 1925, XX4; "Evolution Battle to Go to Congress; New Law is Sought," July 24, 1925, 1; "Professors to War on Evolution Laws," January 1, 1927, 1; "College Professors, Scientists and Fundamentalists Prepare to Do Battle—The Two Sides Presented," January 30, 1927, XX3; "Protest in Florida on Anti-Darwin Bill," April 30, 1927, 3; "Missouri Kills Anti-Evolution Bill," February 9, 1927, 8; "Act on Evolution Bills," February 10, 1927, 38; "Beat Anti-Evolution Bill," April 14, 1927, 24. For a brief summary of state laws, see Richard David Wilhelm, "A Chronology and Analysis of Regulatory Actions Relating to the Teaching of Evolution in Public Schools" (PhD dissertation, University of Texas–Austin, 1978), 61–64. Wilhelm does not include in his list an Iowa bill of 1923 (House File 657), two 1926 Mississippi bills (House Bill 39 and Senate Bill 214), a 1927 West Virginia House Bill (358), or Arkansas' initiative measure of 1928. See also Maynard Shipley, "Growth of the Anti-Evolution Movement," *Current History* XXXII (1930): 330–32. Many historians have accepted Shipley's numbers without question, but he made a few errors. For instance, he assigned Kentucky's and South Carolina's first debates to 1921, when in fact both considered the bills in 1922. He omits one Tennessee bill, considered in 1923, one Iowa bill in 1923, one Georgia bill in 1925, two Louisiana bills in 1926, a South Carolina bill in 1927, and two Alabama bills in 1927. He does not include bills that repeated similar legislation in the same state in the same year. He also asserts a consideration of a resolution in Oklahoma in 1929 and a bill in Kentucky in 1928 for which I can find confirmation in no other source. He includes in his total state executive actions in California and North Carolina. He states that Atlanta, Georgia passed a citywide antievolution rule, when in fact it did not. Atlanta's rule only stipulated that teachers be questioned about their teaching of evolution. Bills, resolutions, or riders known primarily as antievolution measures were introduced in Alabama (SJR 55, 1923; HB 969, 1927; HB 1103, 1927; HB 30, 1927), Arkansas (HB 34, 1927; Act 1, 1928), California (AB 145, 1927), Delaware (HB 92, 1927), Florida (HCR 7, 1923; HB 691, 1925; HB 87, 1927), Georgia (HR 58, 1923; HR 93, 1923; HB 731, 1924; Amendment, 1925), Iowa (HF 657, 1923), Kentucky (HB 260, 1922; HB 191, 1922; SB 136, 1922; HB 96, 1926), Louisiana (HB 41, 1926; HB 208, 1926; HB 279, 1926; HB 314, 1926), Maine (HP 834, 1927), Minnesota (SB 701, 1927; HB 837, 1927), Mississippi (HB 39, 1926; HB 77, 1926; SB 214, 1926),

Missouri (HB 89, 1927), New Hampshire (HB 268, 1927), North Carolina (HR 10, 1925; HB 263, 1927), North Dakota (HB 227, 1927), Oklahoma (HB 197, 1923; HB 81, 1927), South Carolina (Amendment, 1922; HB 60, 1927), Tennessee (SB 681, 1923; HB 947, 1923; SB 133, 1925; HB 185, 1925; HB 252, 1925), Texas (HB 97, 1923; HCR 6, 1923; HB 378, 1925; HB 90, 1929), and West Virginia (HB 153, 1923; HB 175, 1925; Resolution, 1927; HB 264, 1927; HB 358, 1927). Many of these states considered several versions of proposed resolutions or legislation. Evolution-restricting acts were passed in Florida, Oklahoma, Tennessee, Mississippi, and Arkansas. In addition, the U.S. Congress heard at least two antievolution arguments during the 1920s. See Willard B. Gatewood Jr., ed., *Controversy in the Twenties: Fundamentalism, Modernism, and Evolution* (Nashville, TN: Vanderbilt University Press, 1969), 321–29.

13. Alabama House Bill 30, *Journal of the House of Representatives of the State of Alabama 1927*, 89; Alabama House Bill 969, *Journal of the House of Representatives of the State of Alabama 1927*, 1622, 1983, 2040; Alabama House Bill 1103, *Journal of the House of Representatives of the State of Alabama 1927*, 2153, 2426–27, 2596; Arkansas House Bill 34, *Journal of the House of Representatives for the General Assembly of the State of Arkansas 1927*, 68–69, 263, 323–24; California Assembly Bill 145, *Journal of the Assembly during the Forty-Seventh Session of the Legislature of the State of California 1927*, 182, 445, 482, 543–44, 566, 2639–41; New Hampshire House Bill 268, "Journal of the House of Representatives 1927," *Journals New Hampshire Senate and House 1927*, 154, 274; North Dakota House Bill 222, *State of North Dakota Journal of the House of the Twentieth Session of the Legislative Assembly 1927*, 519, 1022; Oklahoma House Bill 81, *Journal of the House of Representatives of the Legislature of the State of Oklahoma 1927*, 281, 305; South Carolina House Bill 60, *Journal of the House of Representatives of the First Session of the 78th General Assembly of the State of South Carolina 1927*, 70, 1128, 1482; West Virginia House Resolution, *Journal of the House of Delegates of West Virginia 1927*, 65, 97–98; West Virginia House Bill 264; *Journal of the House of Delegates of West Virginia 1927*, 104, 663; West Virginia House Bill 358, *Journal of the House of Delegates of West Virginia 1927*, 129; North Carolina House Bill 263, *Journal of the House of Representatives of the General Assembly of the State of North Carolina 1927*, 85, 241; Delaware House Bill 92, *Delaware House Journal 1927*, 156; Maine House Paper 834, *Legislative Record of the Eighty-Third Legislature of the State of Maine 1927*, 239, 242, 247–49, 313, 835; Minnesota Senate Bill 701, *Journal of the Senate of the State of Minnesota 1927*, 508–9; Florida House Bill 87, *Florida House Journal 1927*, 3000–3001; Wilhelm, "Chronology and Analysis," 373 [Missouri House Bill No. 89]. See also Maynard Shipley, "A Year of the Monkey War," *The Independent* 119 (October 1, 1927): 326–45. Shipley left out any mention of South Carolina's House Bill number 60.

14. Moran, *Scopes Trial*, 2; Larson, *Summer for the Gods*, 203, 142.

15. Kenneth K. Bailey, "The Antievolution Crusade of the Nineteen-Twenties" (PhD dissertation, Vanderbilt University, 1954), 68–69, 71, 222–24; Gerald Skoog, "The Coverage of Human Evolution in High School Biology Textbooks in the 20th Century and in Current State Science Standards," *Science and Education* 14 (2005): 398.

16. Kentucky House Bill 191, *Journal of the House of Representatives of the Commonwealth of Kentucky 1922*, 1668–71. See also Wilhelm, "Chronology and Analysis,"

316. Wilhelm's appendices include selections from much of the fundamentalist legislation of the 1920s.

17. "Pastors Condemn Evolution Theory in Public Schools," *Minneapolis Journal*, October 26, 1922.

18. Willard B. Gatewood Jr., *Preachers, Pedagogues and Politicians: The Evolution Controversy in North Carolina, 1920–1927* (Chapel Hill, NC: University of North Carolina Press, 1966), 149.

19. Summers Amendment, 68th Cong., 1st sess., *Congressional Record* 65 (May 3, 1924): H 7796.

20. Gatewood, *Preachers, Pedagogues and Politicians*, 222 [text of Poole Bill, North Carolina House Bill No. 263]; North Carolina House Bill 263, *Journal of the House of Representatives of the General Assembly of the State of North Carolina 1927*, 85, 241.

21. West Virginia House Bill 264; *Journal of the House of Delegates of West Virginia 1927*, 104.

22. Florida House Bill 87, *Florida House Journal* 1927, 3000–3001.

23. "Editorial: Stop and Think," *Louisville Courier-Journal*, February 3, 1922, 2.

24. See George M. Marsden, *Fundamentalism and American Culture, The Shaping of Twentieth-Century Evangelicalism, 1870–1925* (New York: Oxford University Press, 1980); Larson, *Summer for the Gods*; Ronald L. Numbers, *The Creationists: From Scientific Creationism to Intelligent Design* (Cambridge, MA: Harvard University Press, 2006); Ronald L. Numbers, *Darwinism Comes to America* (Cambridge, MA: Harvard University Press, 1999); Michael Lienesch, *In the Beginning: Fundamentalism, the Scopes Trial, and the Making of the Antievolution Movement* (Chapel Hill, NC: University of North Carolina Press, 2007). See also Gatewood, *Preachers, Pedagogues and Politicians*; Wilhelm, "Chronology and Analysis"; Bailey, "The Antievolution Crusade"; LeRoy Johnson, "The Evolution Controversy During the 1920's" (PhD dissertation, New York University, 1954); Virginia Gray, "Anti-Evolution Sentiment and Behavior: The Case of Arkansas," *Journal of American History* 62 (1970): 353–65.

25. Jonathan Zimmerman, *Whose America? Culture Wars in the Public Schools* (Cambridge, MA: Harvard University Press, 2002), 1–8.

Chapter 1

1. Arthur B. Patten, "Mysticism and Fundamentalism," *Christian Century* 40 (March 8, 1923): 297.

2. H. L. Mencken, "Protestantism in the Republic," in *Prejudices: Fifth Series* (New York: Alfred A. Knopf, 1926), 111.

3. Stewart G. Cole, *The History of Fundamentalism* (Westport, CT: Greenwood Press, 1931), 306.

4. H. Richard Niebuhr, "Fundamentalism," in *Encyclopedia of the Social Sciences*, 6 (New York: MacMillan, 1931): 527.

5. Norman F. Furniss, *The Fundamentalist Controversy, 1918–1931* (New Haven, CT: Yale University Press, 1954), 28. See also the popular history by Frederick

Lewis Allen, *Only Yesterday: An Informal History of the Nineteen-Twenties* (New York: Harper and Bros., 1931).

6. George M. Marsden, *Fundamentalism and American Culture: The Shaping of Twentieth-Century Evangelicalism, 1870–1925* (New York: Oxford University Press, 1980), 4; Ernest R. Sandeen, *The Roots of Fundamentalism: British and American Millenarianism, 1800–1930* (Chicago, IL: University of Chicago Press, 1970); Marsden, "Defining Fundamentalism," *Christian Scholar's Review* 1 (Winter 1971): 141–51; Sandeen, "Defining Fundamentalism, A Reply to Professor Marsden," *Christian Scholar's Review* 1 (Spring 1971): 227–32.

7. See, for example, William V. Trollinger, *God's Empire: William Bell Riley and Midwestern Fundamentalism* (Madison, WI: University of Wisconsin Press, 1990); Virginia L. Brereton, *Training God's Army: The American Bible School, 1880–1940* (Bloomington, IN: Indiana University Press, 1990); Joel A. Carpenter, *Revive Us Again: The Reawakening of American Fundamentalism* (New York: Oxford University Press, 1997); Edward Larson, *Summer for the Gods: The Scopes Trial and America's Continuing Debate over Science and Religion* (Cambridge, MA: Harvard University Press, 2001); Ronald L. Numbers, *Darwinism Comes to America* (Cambridge, MA: Harvard University Press, 1999), 76–91; Barry Hankins, *God's Rascal: J. Frank Norris and the Beginnings of Southern Fundamentalism* (Lexington, KY: University of Kentucky Press, 1996); Mark Taylor Dalhouse, *An Island in the Lake of Fire: Bob Jones University, Fundamentalism, and the Separatist Movement* (Athens, GA: University of Georgia Press, 1996); Dale E. Soden, *The Reverend Mark Matthews: An Activist in the Progressive Era* (Seattle: University of Washington Press, 2001); Margaret Lamberts Bendroth, *Fundamentalism and Gender, 1875 to the Present* (New Haven, CT: Yale University Press, 1993); Bendroth, *Fundamentalists in the City: Conflict and Division in Boston's Churches, 1885–1950* (New York: Oxford University Press, 2005); and Jeffrey P. Moran, "The Scopes Trial and Southern Fundamentalism in Black and White: Race, Region, and Religion," *Journal of Southern History* 70 (February 2004): 95–120.

8. See Numbers, *Darwinism Comes to America*.

9. Jon H. Roberts, *Darwinism and the Divine: Protestant Intellectuals and Organic Evolution, 1859–1900* (Madison: University of Wisconsin Press, 1988), 233.

10. Roberts, *Darwinism and the Divine*, 31.

11. Marsden, *Fundamentalism and American Culture*, 17–18.

12. William R. Hutchison, *The Modernist Impulse in American Protestantism* (Cambridge, MA: Harvard University Press, 1976), 2.

13. Marsden, *Fundamentalism and American Culture*, 11–42.

14. Shirley Jackson Case, *The Revelation of John* (Chicago, IL: University of Chicago Press, 1919); Shailer Mathews, *The Faith of Modernism* (New York: MacMillan, 1924).

15. U.S. Department of Commerce, Bureau of the Census, *Historical Statistics of the United States, Colonial Times to 1970, Bicentennial Edition, Part 1* (Washington, DC: U.S. Department of Commerce, Bureau of the Census,1975), 383, 15, 386, 379.

16. "Atlanta School Attendance Records Smashed; Figures Indicate City Population of 320,000," *Atlanta Constitution*, September 13, 1925, 1; "25,000 Children Attend School," *Atlanta Constitution*, September 9, 1919, 8.

17. Lawrence Cremin, *American Education: The Metropolitan Experience, 1876–1980* (New York: Harper & Row, 1988), 544–50.
18. Marsden, *Fundamentalism and American Culture*, 43–140.
19. Sandeen, *The Roots of Fundamentalism*, 69; Marsden, *Fundamentalism and American Culture*, 113.
20. Marsden, *Fundamentalism and American Culture*, 51.
21. Ibid., 46.
22. Ibid., 54. For more on the theology, see Carpenter, *Revive Us Again*, 247–49; see also Tim Weber, *Living in the Shadow of the Second Coming: American Premillennialism, 1875–1982* (Chicago, IL: University of Chicago Press, 1987). For recent popularizations of the endtime prophecies, see Hal Lindsay with C. C. Carlson, *The Late Great Planet Earth* (Grand Rapids, MI: Zondervan Press, 1970); or the more recent series by Tim LaHaye and Jerry Jenkins, *Left Behind*, 12 vols. (Carol Stream, IL: Tyndale House Publishers, 1996–2007).
23. Sandeen, *The Roots of Fundamentalism*, 268.
24. Marsden, *Fundamentalism and American Culture*, 73. See also Melvin Dieter, *The Holiness Revival of the Nineteenth Century* (Metuchen, NJ: Scarecrow Press, 1980); and Timothy L. Smith, *Called Unto Holiness: The Story of the Nazarenes, the Formative Years* (Kansas City, MO: Nazarene Publishing House, 1962).
25. Marsden, *Fundamentalism and American Culture*, 102.
26. Ibid., 125–38.
27. Daniel T. Rodgers, *Contested Truths: Keywords in American Politics since Independence* (New York: Basic Books, 1987), 11.
28. R. A. Torrey and A. C. Dixon, ed., *The Fundamentals: A Testimony to the Truth* (Chicago, IL: Testimony Publishing, 1910–1915); Marsden, *Fundamentalism and American Culture*, 118–19.
29. Curtis Lee Laws, "Fundamentalism from the Baptist Viewpoint," *Moody Bible Institute Monthly* [*Moody Monthly*] 23 (September 1922): 1, 15.
30. Harry Emerson Fosdick, "Shall the Fundamentalists Win?" *Christian Century*, 39 (June 12, 1922): 713–17.
31. Soden, *The Reverend Mark Matthews*, 171–72; Marsden, *Fundamentalism and American Culture*, 173–74; Sandeen, *The Roots of Fundamentalism*, 252.
32. Ibid., 173; Marsden, *Fundamentalism and American Culture*, 180.
33. Trollinger, *God's Empire*, 55–57.
34. Ibid., 56.
35. William Bell Riley, "Report of Committee on Resolutions," in *God Hath Spoken* (Philadelphia: Bible Conference Committee, 1919), 13.
36. Charles A. Blanchard, "Report of Committee on Correlation of Colleges, Seminaries and Academies," *God Hath Spoken*, 19–20.
37. Lawrence W. Levine, *Defender of the Faith: William Jennings Bryan: The Last Decade, 1915–1925* (New York: Oxford University Press, 1965), 264; Larson, *Summer for the Gods*, 41.
38. William Jennings Bryan, *The Bible and Its Enemies* (Chicago, IL: Bible Institute Colportage Association, 1921), 33–34.

39. Ferenc M. Szasz, "Three Fundamentalist Leaders: The Roles of William Bell Riley, John Roach Straton, and William Jennings Bryan in the Fundamentalist-Modernist Controversy" (PhD dissertation, University of Rochester, 1969), 148.

40. Olive C. Wadlin to John Roach Straton, June 3, 1922; Straton to Olive C. Wadlin, June 8, 1922, John Roach Straton Papers, American Baptist—Samuel Colgate Historical Library; American Baptist Historical Society, Rochester, NY.

41. Quoted in Larson, *Summer for the Gods*, 55; *Memphis Commercial Appeal*, February 1925, quoted in Larson, *Summer for the Gods*, 54–55.

42. William G. McLoughlin Jr., *Billy Sunday Was His Real Name* (Chicago, IL: University of Chicago Press, 1955), 260, 270, 283; Robert F. Martin, *Hero of the Heartland: Billy Sunday and the Transformation of American Society, 1862–1935* (Bloomington, IN: Indiana University Press, 2002), 122–34.

43. Soden, *The Reverend Mark Matthews*, 179.

44. Ibid., 188.

45. "Editorial: Stop and Think," *Louisville Courier-Journal*, February 3, 1922.

46. "Special Report: Darwinian Theory Stirs Up Kentucky," *New York Times*, February 2, 1922, 11.

47. J. H. Ralston, "Notes and Suggestions," *Moody Monthly* 21 (February 1921): 268.

48. William Jennings Bryan, "God and Evolution," *New York Times*, February 26, 1922, 1, 11.

49. Bryan, *The Bible and Its Enemies*, 23.

50. John Roach Straton, "Sinclair Lewis's 'Elmer Gantry,'" box 10, file "Lewis, Sinclair," Straton Papers.

Chapter 2

1. Samuel Zane Batten, "The Battle Within the Churches," *The Searchlight* 6 (October 26, 1923): 1. Batten was the chairman of the Social Service Committee of the Northern Baptist Convention.

2. Thomas F. Gieryn, "Boundaries of Science," in *Handbook of Science and Technology Studies*, ed. Sheila Jasanoff (Thousand Oaks, CA: Sage Publications, 1994), 405.

3. Syntheses of recent work on identity include Kwame Anthony Appiah, *The Ethics of Identity* (Princeton, NJ: Princeton University Press, 2005); Amartya Sen, *Identity and Violence: The Illusion of Destiny* (New York: Norton, 2006). For brief guides to the sociological literature on identity, see Karen A. Cerulo, "Identity Construction: New Issues, New Directions," *Annual Review of Sociology* 23 (1997): 385–409; and Francesca Polletta and James M. Jasper, "Collective Identity and Social Movements," *Annual Review of Sociology* 27 (2001): 283–305. See also Michael Lienesch, *In the Beginning: Fundamentalism, the Scopes Trial, and the Making of the Antievolution Movement* (Chapel Hill, NC: University of North Carolina Press, 2007). Lienesch explores the antievolution movement as an enduring social and cultural identity.

4. See William Vance Trollinger Jr., *God's Empire: William Bell Riley and Midwestern Fundamentalism* (Madison, WI: University of Wisconsin Press, 1990).

5. See, for example, his defenses of "orthodoxy": William Bell Riley, "The Challenge of Orthodoxy," *Christian Fundamentals in School and Church* [*CFSC*] 2 (July–September

1920): 365–68; William Bell Riley, "Address to Denver Conference on Christian Fundamentals," *CFSC* 3 (July–September 1921): 6.

6. See *CFSC* 5 (January–March 1923), 6 (October–December 1923).

7. William Bell Riley, "Questionnaire," *CFSC* 6 (October–December 1923): 8–11.

8. Ibid., 11.

9. William Bell Riley, "Fundamentalism—The Word that has Won Its Way," *CFSC* 6 (October–December 1923): 13.

10. Richard S. Beal, "Fundamentalism: A Call Back to the Bible," *CFSC* 6 (October–December 1923): 30.

11. Riley reprinted this creed many times in pamphlets and in his quarterly publication. See also Trollinger, *God's Empire*, 163.

12. William Bell Riley, "Notes," *CFSC* 5 (April–June 1923): 24.

13. Barry Hankins, *God's Rascal: J. Frank Norris and the Beginnings of Southern Fundamentalism* (Lexington, KY: University Press of Kentucky, 1996); Trollinger, *God's Empire*, 43.

14. J. E. Conant, "Can Northern Baptists Stay Together?" *Searchlight* 4 (April 4, 1922): 22.

15. J. Frank Norris, "The Alabama Baptist Attacks Fundamentalists and Defends Evolution," *Searchlight* 6 (February 16, 1923): 1–5.

16. J. Frank Norris, "World's Fundamentals Convention," *Searchlight* 6 (March 16, 1923): 1.

17. J. Frank Norris, "Letter to the Editor of the *Dallas News*," *Searchlight* 7 (January 25, 1924): 1; J. Frank Norris, "The Battle Within the Churches," *Searchlight* 6 (October 26, 1923): 1.

18. "J. Frank Norris Kills Ft. Worth Lumber Dealer in Church; Pleads Self Defense," *Waco News-Tribune*, July 18, 1926.

19. Virginia L. Brereton, *Training God's Army: The American Bible School, 1880–1940* (Bloomington, IN: Indiana University Press, 1990), 79–84; Gene A. Getz, *MBI: The Story of Moody Bible Institute* (Chicago, IL: Moody Press, 1969); Bernard R. De Remer, *Moody Bible Institute: A Pictorial History* (Chicago, IL: Moody Press, 1960).

20. James M. Gray, "Editorial Notes," *Moody Bible Institute Monthly* [*Moody Monthly*] 21 (April 1921): 347; James M. Gray, "Editorial Notes," *Moody Monthly* 22 (May 1922): 1003.

21. James M. Gray, "Editorial Notes," *Moody Monthly* 23 (September 1922): 1, 4.

22. See Brereton, *Training God's Army*, 79–84; and Robert Williams, *Chartered for His Glory: Biola University, 1908–1983* (La Mirada, CA: Associated Students of Biola University, 1983).

23. B. W. Burleigh, "An Open Letter to a Modernist," *The King's Business* 14 (March 1923): 244–45.

24. See Michael Kazin, *A Godly Hero: The Life of William Jennings Bryan* (New York: Anchor Books, 2007); Lawrence W. Levine's *Defender of the Faith: William Jennings Bryan: The Last Decade, 1915–1925* (New York: Oxford University Press, 1965).

25. Frederick F. Shannon, "Bryanism," *The Christian Century* 39 (April 6, 1922): 428–31.

26. Glenn Frank, "William Jennings Bryan: A Mind Divided Against Itself," *The Christian Century* 106 (September 1923): 794.

27. Anonymous editorial in *King's Business* 14 (July–August, 1923): 805.

28. L. M. Aldridge, "Editorial," *Searchlight* 8 (August 7, 1925): 1.

29. William Bell Riley, "Notes," *CFSC* 5 (April–June 1923): 24. See also Levine, *Defender of the Faith*, 264.

30. William Jennings Bryan, "Misrepresentations of Darwinism and Its Disciples," *Moody Monthly* 23 (April 1923): 331.

31. William Jennings Bryan, *Bryan's Last Speech: Undelivered Speech to the Jury in the Scopes Trial* (Oklahoma City, OK: Sunlight Publishing Society, 1925), 46.

32. Bryan to William Bell Riley, 7 June 1925, Bryan Papers, Library of Congress, Washington DC.

33. Kazin, *A Godly Hero*, 264.

34. William Jennings Bryan, "Letter to the Editor," *Chicago Daily Tribune*, May 28, 1925, quoted in Levine, *Defender of the Faith*, 274.

35. Kazin, *A Godly Hero*, 45–80.

36. C. Allyn Russell, *Voices of American Fundamentalism: Seven Biographical Studies* (Philadelphia: Westminster Press, 1976), 148.

37. J. Gresham Machen, "Statement" unpublished typescript, 1926 folder, box 2 of J. Gresham Machen writings, Machen Papers, Montgomery Library archive of Westminster Theological Seminary, Philadelphia, PA. The writings describe Machen's stance on Prohibition.

38. D. G. Hart, *Defending the Faith: J. Gresham Machen and the Crisis of Conservative Protestantism in Modern America* (Baltimore, MD: The Johns Hopkins University Press, 1994), 68.

39. J. Gresham Machen, *Christianity and Liberalism* (Grand Rapids, MI: Eerdmans Publishing Co., 1923).

40. J. Gresham Machen and Charles P. Fagnani, "Does Fundamentalism Obstruct Social Progress?" *Survey Graphic* 5 (July 1924): 389–92, 425–27.

41. J. Gresham Machen, "What Fundamentalism Now Stands For," *New York Times*, June 21, 1925, XX1.

42. Machen to Albert Sydney Johnson, 23 May 1925, Machen Papers.

43. Machen to Albert Sydney Johnson, 30 June 1925, Machen Papers.

44. William Bell Riley, "An Orthodox Premillennial Seminary," *CFSC* 5 (January–March 1923): 20; William Bell Riley, "The Great Objective of the Fort Worth Convention," *CFSC* 4 (April–June 1923): 25–26.

45. John D. Hannah, "Social and Intellectual Origins of the Evangelical Theological College" (PhD dissertation, University of Texas at Dallas, 1988), 212–13, 174–90. Quotation is from page 174.

46. Hannah, "Social and Intellectual Origins of the Evangelical Theological College," 194–95.

47. *Evangelical Theological College Bulletin* 12 (July–September 1936): 2.

48. "Our New Home," *Evangelical Theological College Bulletin* 2 (August 1926): 3.

49. *Evangelical Theological College Bulletin* 2 (June 1926): 6–7; Rudolf Albert Renfer, "A History of Dallas Theological Seminary" (PhD disertation, University of Texas–Austin, 1959), 314.

50. Timothy Weber, *Living in the Shadow of the Second Coming: American Premillennialism, 1875–1982* (Chicago, IL: University of Chicago, 1987), 238.

51. Lewis Sperry Chafer to A. C. Gaebelein, November 13, 1923; President's Office Papers, Accession 2006-18, L. S. Chafer correspondence, box 12, file 23, Archives, Dallas Theological Seminary.

52. A. C. Gaebelein to Chafer, December 21, 1923; President's Office Papers, Accession 2006–22, Unprocessed Chafer Papers, box 1, file "Correspondence concerning founding and location of Seminary," Archives, Dallas Theological Seminary.

53. Chafer to Gaebelein, November 13, 1923; President's Office Papers, Accession 2006-18, L. S. Chafer correspondence, box 12, file 23, Archives, Dallas Theological Seminary;

54. Ibid.

55. W. H. Griffith Thomas, "Fundamentalism and Modernism: Two Religions," *Christian Century* 41 (January 3, 1924): 5–6, quoted in Renfer, "A History of Dallas Theological Seminary," 39.

56. See Grant Wacker, *Heaven Below: Early Pentecostals and American Culture* (Cambridge, MA: Harvard University Press, 2001).

57. Grant Wacker, "Travail of a Broken Family: Radical Evangelical Responses to the Emergence of Pentecostalism in America, 1906–16," in *Pentecostal Currents in American Protestantism*, ed. Edith Blumhofer, Russell P. Spittler, and Grant A. Wacker (Urbana, IL: University of Illinois Press, 1999), 23–50.

58. J. Frank Norris, "Pentecostal Preacher 'Fleeces' Flock," *Searchlight* 7 (May 2, 1924): 1; J. Frank Norris, "J. Frank Norris on 'Speaking with Tongues,'" *Searchlight* 7 (May 9, 1924): 4.

59. Wacker, *Heaven Below*, 145. See also Edith L. Blumhofer, *Aimee Semple McPherson: Everybody's Sister* (Grand Rapids, MI: Eerdmans Publishing Co., 1993), 2.

60. 'A Former Sympathizer of the Pentecostal Movement,' "Pentecostal Saints and the Tongues Movement," *Moody Monthly* 21 (January 1921): 211.

61. James M. Gray, "What About Mrs. Aimee Semple McPherson," *Moody Monthly* 22 (November 1921): 649.

62. Amzi C. Dixon, "Speaking in Tongues," *King's Business* 13 (January 1922): 14–17.

63. C. F. Koehler, "What the Bible Says About Speaking in Tongues," *Moody Monthly* 22 (February 1922): 808.

64. Keith Brooks, "Dangerous Methods of Seeking the Holy Spirit's Power" *Moody Monthly* 24 (October 1923): 55.

65. Arno C. Gaebelein, "Christianity vs. Modern Cults," *Moody Monthly* 22 (March 1922): 858.

66. Gary B. Ferngren, "The Evangelical-Fundamentalist Tradition," in *Caring and Curing*, ed. Ronald Numbers and Darrel Amundsen (Baltimore, MD: The Johns Hopkins Press, 1998), 486–513; William Bell Riley, prefatory note to C. I. Scofield, "How God Heals Sickness," *CFSC* 3 (April–June 1921): 28.

67. J. Frank Norris, "Rev. J. Frank Norris Preaches on Divine Healing and Effect," *Fort Worth Record*, repr. in *Searchlight* 2 (April 1, 1920): 2; J. Frank Norris, "Scriptural View on Healing," *Searchlight* 2 (April 8, 1920); A. T. Pierson, "What About Divine Healing? Have Supernatural Signs Ceased during the Church Age?" *The King's Business* 12 (March 1921): 231.

68. Quoted in Milton L. Rudnick, *Fundamentalism and the Missouri Synod: a Historical Study of their Interaction and Mutual Influence* (St. Louis, MO: Concordia Publishing, 1966), 79.

69. James M. Gray, "Editorial Notes," *Moody Monthly* 22 (June 1922): 1052.

70. Jessie Sage Robertson, "'The Spirit of Truth and the Spirit of Error': Eye-Opening Testimonies Regarding Cults—Seventh Day Adventism," *King's Business* (September 1920): 856.

71. S. Fraser Langford, "Prophetic Mistakes of the Adventists," *King's Business* (February 1923): 126.

72. Grant Stroh, "Practical and Perplexing Questions," *Moody Monthly* 22 (September 1921): 574.

73. James M. Gray, from a review of I. M. Haldeman's *Truth About the Sabbath and the Lord's Day*, *Moody Monthly* 21 (November 1921): 137.

74. A. R. Funderburk, "The Ku Klux Klan—Is It of God?" *Moody Monthly* 23 (March 1923): 291–92; John Bradbury, "Defending the Ku Klux Klan—A Reply to Mr. Funderburk," *Moody Monthly* 23 (May 1923): 420–21.

75. James M. Gray, "Editorial," *Moody Monthly* 23 (February 1923): 240; James M. Gray, "Editorial Notes," *Moody Monthly* 24 (December 1923): 163.

76. J. Frank Norris, "Judge Wilson, K.C.'s, Ku Klux Klan and Bootleggers," *Searchlight* 4 (May 22, 1922): 1; J. Frank Norris, "The Menace of Roman Catholicism in Politics," *Searchlight* (August 1, 1924): 1. See also the advertisements for Klan publications and events, *Searchlight* (September 15, 1922): 4; (March 14, 1924): 3; (December 7, 1924): 3. For his campaign against Governor Ferguson, see especially his *Searchlight* of August 15, 1924. See also Shelley Sallee, "'The Woman of It': Governor Miriam Ferguson's 1924 Election." *Southwestern Historical Quarterly* 100, no. 1 (1996): 1–16. For the relationship between religious conservatives and other conservatives in the 1930s and beyond, see Leo P. Ribuffo, *The Old Christian Right: The Protestant Far Right from the Great Depression to the Cold War* (Philadelphia: Temple University Press, 1983). Ribuffo's point about the 1930s is equally applicable to the 1920s: "Within the United States during the 1930s . . . not all bigots were fundamentalists, and not all fundamentalists were bigots" (249).

77. "Man Was Never a Jelly Fish, Declares Writer in Attack on Evolution Theory," *Imperial Nighthawk* 1 (May 16, 1923): 2; "Donahey Defies Klan in Vetoing Bible Measure," *Cleveland Plain Dealer*, May 1, 1925; Leonard J. Moore, *Citizen Klansmen: The Ku Klux Klan in Indiana, 1921–1928* (Chapel Hill, NC: University of North Carolina Press), 41–44.

78. Wacker, *Heaven Below*, 178–84.

79. Ibid., 182.

80. Ibid., 75.

81. Edith L. Blumhofer, *Restoring the Faith: The Assemblies of God, Pentecostalism, and American Culture* (Urbana, IL: University of Illinois Press, 1993), 160.

82. For one example of a prominent Seventh-day Adventist who considered himself, and all Adventists, to be fundamentalists, see William H. Branson, *Reply to Canright: The Truth About Seventh-day Adventists* (Washington, DC: Review and Herald Publishing Association, 1933), 22. For a concise historical view of the debate about fundamentalism within Seventh-day Adventism, see Gary Land, "Shaping the Modern Church 1906–1930," in *Adventism in America: A History* (Grand Rapids, MI: Eerdmans Publishing Co., 1986), 139–69. I am indebted to Ronald Numbers for these references. For George McCready Price's attitude toward fundamentalism, as well as the attitudes of several leading fundamentalists of the

1920s toward Price, see Ronald Numbers, *The Creationists* (1993), 96–100. The *Science* editor is quoted on page 73.

83. Susie C. Stanley, "Wesleyan/Holiness Churches: Innocent Bystanders in the Fundamentalist/Modernism Controversy," in *Re-Forming the Center: American Protestantism, 1900 to the Present*, ed. Douglas Jacobsen and William Vance Trollinger Jr. (Grand Rapids, MI: Eerdmans Publishing Co., 1998), 172–93.

84. W. B. McCreary, "How Long Go Ye Limping?" *Gospel Trumpet* 38 (April 3, 1924): 6.

85. Maynard Shipley, *The War on Modern Science* (New York: Alfred A. Knopf, 1927), 47.

86. *Moody Monthly* 25 (April 1925): 371–72.

87. Hankins, *God's Rascal*, 40–41.

88. George M. Marsden, *Fundamentalism and American Culture, The Shaping of Twentieth-Century Evangelicalism, 1870–1925* (New York: Oxford University Press, 1980),180–84.

89. Quoted in Blumhofer, *Restoring the Faith*, 160.

90. Matthew Avery Sutton, *Aimee Semple McPherson and the Resurrection of Christian America* (Cambridge, MA: Harvard University Press, 2007), 37, 113, 120–22, 216.

91. Fred "Fritz" L. Harper to Machen, 4 May 1925, Machen Papers; Albert Sidney Johnson to Machen, 19 May 1925, Machen Papers; Machen to Albert Sidney Johnson, 23 May 1925, Machen Papers; Machen to Mark A. Matthews, 4 December 1924, Machen Papers; Frederick Erdman to Leander S. Keyser, 24 March 1928, Machen Papers; Dale E. Soden, *The Reverend Mark Matthews: An Activist in the Progressive Era* (Seattle, WA: University of Washington Press, 2001), 174–75.

Chapter 3

1. Lawrence A. Cremin, *American Education: The Metropolitan Experience, 1876–1980* (New York: Harper & Row, 1988), 12.

2. James L. Axtell, "The School Upon a Hill: Education and Society," in *Colonial New England* (New Haven, CT: Yale University Press, 1985), 285.

3. Benjamin Justice, *The War that Wasn't: Religious Conflict and Compromise in the Common Schools of New York State, 1865–1900* (Albany, NY: SUNY Press, 2005), 2–10. See also James Davison Hunter, *Culture Wars: The Struggle to Define America* (New York: Basic Books, 1991) and Jonathan Zimmerman, *Whose America? Culture Wars in the Public Schools* (Cambridge, MA: Harvard University Press, 2002).

4. William Jennings Bryan to W. J. Singleterry, 11 April 1923, Bryan Papers, Library of Congress, Washington, DC; Bryan, "Letter to the Editor of the *Chicago Tribune*," June 14, 1923, Bryan Papers; Bryan to C. H. Thurber, 22 December 1923, Bryan Papers.

5. Alfred Fairhurst, *Atheism in Our Universities* (Cincinnati, OH: Standard Publishing Co., 1923), 84.

6. Ibid., 92. Emphasis in original.

7. T. T. Martin, *The Evolution Issue* (Los Angeles, CA: n.p., 1923?), 38–39.

8. James M. Gray, "The Sacred Cow of Evolution," *Moody Bible Institute Monthly* [*Moody Monthly*] 29 (January 1929): 225.

9. William Bell Riley, "Higher Criticism in College Libraries," *School and Church* 2 (January–March 1920): 298.

10. Martin, *The Evolution Issue*, 17.

11. Merle Curti, *The Growth of American Thought* (New York: Harper and Row, 1964), 529.

12. George M. Marsden, *The Soul of the American University: From Protestant Establishment to Established Nonbelief* (New York: Oxford University Press, 1994).

13. Quoted in Marsden, *Soul of the American University*, 267.

14. Richard Sutch & Susan B. Carter, ed., *Historical Statistics of the United States, Millennial Edition* (New York: Cambridge University Press, 2006), http://hsus.cambridge .org.ezproxy.library.wisc.edu/HSUSWeb/ (accessed July 21, 2006).

15. Marsden, *Soul of the American University*, 123–33. The quote is from page 127.

16. Ibid., 167–80.

17. Thomas A. Askew Jr. "The Liberal Arts College Encounters Intellectual Change: A Comparative Study of Education at Knox and Wheaton Colleges, 1837–1925" (PhD dissertation, Northwestern University, 1969), 228, 234, 248; Frances Carothers Blanchard, *The Life of Charles Albert Blanchard* (New York: Fleming H. Revell Co., 1932), 182. See also Timothy L. Smith, "Introduction: Christian Colleges and American Culture," in *Making Higher Education Christian: The History and Mission of Evangelical Colleges in America*, ed. Joel A. Carpenter and Kenneth W. Shipps (Grand Rapids: MI: Eerdmans Publishing Co., 1987), 1–2; and William C. Ringenberg, *The Christian College: A History of Protestant Higher Education in America* (Grand Rapids, MI: Eerdmans Publishing Co. and Christian University Press, 1984); "Platform of Wheaton College, Wheaton, Illinois," 3 March 1926, typescript copy in President's Office Papers, Accession 2006-18, L. S. Chafer correspondence, box 10, folder 26, Archives, Dallas Theological Seminary.

18. Kermit L. Staggers, "Reuben A. Torrey: American Fundamentalist, 1856–1928" (PhD dissertation, Claremont Graduate School, 1986), 55.

19. Willard B. Gatewood Jr., *Preachers, Pedagogues and Politicians: The Evolution Controversy in North Carolina, 1920–1927* (Chapel Hill, NC: University of North Carolina Press, 1966), 87.

20. Marsden, *Soul of the American University*, 280–81; see also Jon H. Roberts, "Conservative Evangelicals and Science Education in American Colleges and Universities, 1890–1940," *The Journal of the Historical Society* 5 (September 2005): 297–329.

21. *Alabama Baptist*, September 12, 1906, quoted in James Clyde Harper, "A Study of Alabama Baptist Higher Education and Fundamentalism, 1890–1930" (PhD dissertation, University of Alabama, 1977), 72.

22. T. C. Horton, "Faith-Wrecking Institutions," *King's Business* (February 1919): 100.

23. 'Chicago Pastor,' "Letter to the Editor," *Moody Monthly* 21 (October 1920): 53.

24. William Bell Riley, "Modernism in Baptist Schools," *Christian Fundamentals in School and Church* [*CFSC*] 3 (October–December 1920): 411.

25. William Bell Riley, *The Menace of Modernism* (New York: Christian Alliance, 1917), 76. See also William V. Trollinger, *God's Empire: William Bell Riley and Midwestern Fundamentalism* (Madison, WI: University of Wisconsin Press, 1990), 34.

26. W. H. Griffith Thomas, "Report of Committee on Resolutions," *God Hath Spoken* (Philadelphia: Bible Conference Committee, 1919), 13.
27. T. T. Martin, *Hell and the High School: Christ or Evolution, Which?* (Kansas City, MO: Western Baptist Publishing Co., 1923), 48, 155.
28. James H. Leuba, *The Belief in God and Immortality: A Psychological, Anthropological and Statistical Study*, 2nd ed. (Chicago, IL: The Open Court Publishing Co., 1921); William Jennings Bryan, *The Menace of Darwinism* (Louisville, KY: Pentecostal Publishing Co., 1919?), 32; James H. Leuba to Bryan, 16 May 1925, Bryan Papers.
29. Charles A. Blanchard Papers, Wheaton College Archives, Buswell Library, box 2, file: National Survey of Attitudes of Colleges Toward Certain Evangelical Policies, 1919.
30. Charles A. Blanchard, "Report of Committee on Correlation of Colleges, Seminaries, and Academies," *God Hath Spoken*, 19–20.
31. George Wilson McPherson, *The Crisis in Church and College* (Yonkers, NY: Yonkers Book Co., 1919). See also G. W. McPherson, "What Shall We Do About It?" *King's Business* (February 1920): 151–55.
32. Alfred Fairhurst, *Atheism in Our Universities* (Cincinnati, OH: Standard Publishing Co., 1923), 105.
33. Ibid., 32
34. Ibid., 187–89.
35. Blanchard, "Report of Committee," 19.
36. McPherson, "What Shall We Do About It?" 155.
37. T. C. Horton, "Editorial," *King's Business* (July–August 1923): 788.
38. T. T. Martin, "Glorious Texas," *CFSC* (January–March 1923): 58–59.
39. Edward A. Birge to Charles Henry Caton, 26 October 1921, Edward A. Birge Papers, Wisconsin Historical Society Archives, Madison, Wisconsin.
40. Irvin G. Wyllie, "Bryan, Birge, and the Wisconsin Evolution Controversy, 1921–1922," *Wisconsin Magazine of History* 35 (1951–52): 294–301.
41. William Jennings Bryan, "The Modern Arena," *The Commoner* (June, 1921): 3.
42. Bryan, letter to the editor of the *Chicago Evening Post*, repr. in *The Commoner* (February, 1922): 5; Wyllie, "Bryan, Birge," 299.
43. "Head of U. of W. Atheistic: Bryan," *Chicago Tribune*, September 17, 1921, 1
44. William Jennings Bryan, "Dr. Birge, Autocrat," *The Commoner* (May, 1922): 1.
45. "Bryan Scores Foes of Bible in Address Here," *Milwaukee Daily Journal*, April 17, 1922, 13.
46. Birge to Edwin G. Conklin, 4 March 1922, Birge Papers.
47. Bryan, "Dr. Birge, Autocrat," 1.
48. Ferenc M. Szasz, "William B. Riley and the Fight Against Teaching of Evolution in Minnesota," *Minnesota History* (Spring 1969): 204.
49. "Pastors Condemn Evolution Theory in Public Schools," *Minneapolis Journal*, October 26, 1922, Riley Papers, Billy Graham Center Archives [microform]. See also *CFSC* (January–March 1923): 16.
50. William Bell Riley, "Reply to University Regents on the Evolutionary Controversy," *CFSC* (July–September 1923): 51.
51. Gatewood, *Preachers, Pedagogues, and Politicians*, 111–12.
52. Ibid., 50–56.

53. Curtis Lee Laws, "What Have We a Right to Expect from our Schools?" *Watchman-Examiner* (October 12, 1922): 1293.
54. Virginia L. Brereton, *Training God's Army: The American Bible School, 1880–1940* (Bloomington: Indiana University Press, 1990), 83.
55. Charles A. Blanchard, "Questionnaire for possible teachers," and "Fundamentals of the Christian Faith," [April 15, 1923?] Blanchard papers, Wheaton College Archives, Buswell Library, box 2, file: Student Recruitment.
56. Suzanne Cameron Linder, "William Louis Poteat and the Evolution Controversy," *North Carolina Historical Review* 40 (Spring 1963): 135–57.
57. Gatewood, *Preachers, Pedagogues and Politicians*, 32–34.
58. Ibid., 37–75.
59. Mark A. Greene, "The Baptist Fundamentalists' Case Against Carleton, 1926–1928," *Minnesota History* (Spring 1990): 21.
60. William Bell Riley, "White-Washing Infidelity," *CFSC* (January–March 1922): 7–8.
61. J. Frank Norris, "The Inspiration of the Scriptures: Sermon by the Pastor in Answer to a Book of Higher Criticism by Dr. John A. Rice of Southern Methodist University," *The Searchlight* (May 21, 1921): 1.
62. *CFSC* (July–September 1923): 73.
63. Ibid., 79.
64. Ibid., 77.
65. T. C. Horton, "A Famous Trial," *King's Business* (July–September 1923): 790.
66. Charles Gallaudet Trumbull, "Fundamentalists Expose Modernism in the South," *Searchlight* (May 25, 1923): 1.
67. Lewis Sperry Chafer to A. C. Gaebelein, 2 July 1923, President's Office Papers, Accession 2006-18, L. S. Chafer correspondence, box 12, file 23, Archives, Dallas Theological Seminary.
68. J. Frank Norris, "John A. Rice, the Infidel, Must Go," *Searchlight* (September 23, 1921): 1.
69. J. Frank Norris, "Infidelity in Baylor University," *Searchlight* (October 21, 1921): 1; J. Frank Norris, "Professor Dow and Baylor University," *Searchlight* (November 4, 1921): 1; see also Barry Hankins, *God's Rascal: J. Frank Norris & the Beginnings of Southern Fundamentalism* (Lexington, KY: University Press of Kentucky, 1996), 28.
70. Martin, "Glorious Texas," 58–59; Hankins, *God's Rascal*, 28.
71. John D. Hannah, "The Social and Intellectual Origins of the Evangelical Theological College" (PhD dissertation, University of Texas at Dallas, 1988), 216.
72. Ibid., 218.
73. J. Frank Norris, "'There is Evolution at Baylor' Declared Rev. Jesse Yelvington, a Baylor Student, Before the Baptist Convention," *Searchlight* (December 24, 1922): 1.
74. J. Frank Norris, "Dr. Brooks Defends Professor Sendon," *Searchlight* (December 8, 1922): 1; J. Frank Norris, "Evolution is Now Taught in Baylor University: At This Present Year, Month, Week, Day, Hour, Minute, Second," *Searchlight* (October 12, 1923): 1.
75. J. Frank Norris, "Editorial," *Searchlight* (June 29, 1923): 1; J. Frank Norris, "Editorial," *Searchlight* (November 7, 1924): 1; Barry Hankins, "The Strange Career

of J. Frank Norris: Or, Can a Baptist Democrat Be a Fundamentalist Republican?"
Church History 61 (September 1992): 373–92.
76. J. Frank Norris, "Great Rejoicing! 100% of Baylor Faculty Sign Creedal Statement Which is 100% for Fundamentalism," *Searchlight* (November 28, 1924): 1.
77. Brooks to Mrs. C. L. Sheffield, 20 November 1928, Norris Papers, Billy Graham Center Archives [microform] 2623. Norris's papers also include copies of his incessant correspondence to Brooks and the Baylor board of trustees, demanding the firing of Brooks, the dismissal of certain members of the faculty, and a public review of Baylor's financial records. See, for example, letters 1805, 1807, 2617, 2620, and 2622, Norris papers.
78. Marsden, *Soul of the American University*, 4.
79. Bob Jones, Sr., *The Perils of America, or, Where Are We Headed?* (n.p., n.d.), 24.
80. 'Chicago Pastor,' "Letter to the Editor," 53.

Chapter 4

1. Frank L. McVey, "Address to the People of Kentucky," *Journal of the Kentucky Senate 1922*, 1030–34.
2. "Brands Darwin Bill Foes Apes," *Louisville Courier-Journal*, February 15, 1922, 2.
3. See, for example, "Anti-Darwin Campaigns Stir South and West," *New York Times*, June 10, 1923, X2.
4. William Jennings Bryan, *William Jennings Bryan's Last Speech: Undelivered Speech to the Jury in the Scopes Trial* (Oklahoma City, OK: Sunlight Publishing Society, 1925), 46.
5. William Jennings Bryan, *The Bible and Its Enemies: An Address Delivered at the Moody Bible Institute of Chicago* (Chicago, IL: Bible Institute Colportage Association, 1921), 19.
6. William Jennings Bryan, *The Menace of Darwinism* (Louisville, KY: Pentecostal Publishing Co., n.d.), 4.
7. Harold W. Fairbanks, *Home Geography*, rev. ed. (New York: Educational Publishing Company, 1924), 124.
8. T. T. Martin, *The Evolution Issue* (Los Angeles, CA: n.p., 1923?), 34.
9. Alfred Fairhurst, *Atheism in Our Universities* (Cincinnati, OH: Standard Publishing Co., 1923), 30.
10. William Bell Riley, "Reply to University Regents on the Evolutionary Controversy," *Christian Fundamentals in School and Church* [*CFSC*] 5 (July–September 1923): 51.
11. R. J. Alderman, "Evolution Leads to Sodom," *Monthly Bible Institute Monthly* [*Moody Monthly*] 23 (September 1922): 12.
12. Mary Balch Women's Christian Temperance Union to Bryan, 26 May 1925, Bryan Papers, Library of Congress, Washington, DC. See also W. F. Garvin to Bryan, 19 May 1925, Bryan Papers.
13. J. T. Stroder to Bryan, 30 January 1924, Bryan Papers.
14. "Senate Passes Evolution Bill," *Nashville Banner*, March 13, 1925, 10.
15. Kentucky House Bill 191, *Journal of the House of Representatives of the Commonwealth of Kentucky 1922*, 279, 1668–71; Kentucky Senate Bill 136, *Journal of the*

Kentucky Senate 1922, 891–92, 931, 958, 1043–46, 1061–64, 1082–86; Kentucky House Bill 260, *Journal of the House of Representatives of the Commonwealth of Kentucky 1922*, 414, 587.

16. Richard David Wilhelm, "A Chronology and Analysis of Regulatory Actions Relating to the Teaching of Evolution in Public Schools" (PhD dissertation, University of Texas–Austin, 1978), 323.

17. Georgia House Resolution 58.246-B, *Journal of the House of Representatives of the State of Georgia 1923*, 353–54; Georgia House Resolution 93.390-C, *Journal of the House of Representatives of the State of Georgia 1923*, 553; Georgia House Resolution 93, *Journal of the House of Representatives of the State of Georgia 1923*, 1001–2.

18. Wilhelm, "Chronology and Analysis," 324–25.

19. Texas House Concurrent Resolution 6, *Journal of the House of Representatives of the Third Called Session of the Thirty-Eighth Legislature of Texas 1923*, 73–74, 83.

20. West Virginia House Bill 153, *Journal of the House of Delegates of the State of West Virginia 1923*, 743, 901, 1947; Alabama Senate Joint Resolution 55, *Alabama Senate Journal 1923*, 1211–12; Iowa House File 657, *State of Iowa 1923 Journal of the House*, 758, 1048; Tennessee House Bill 947, *Tennessee House Journal 1923*, 666, 694, 719.

21. Florida House Concurrent Resolution 7, *Journal of the House of Representatives of the State of Florida 1923*, 482–83, 1176, 1853–54, 1878, 2025–27, 2200–2201, 2320, 3187–90.

22. Oklahoma House Bill 197, *Oklahoma House Journal 1923*, 304–5; Oklahoma House Bill 197, *Oklahoma Senate Journal 1923*, 1718–20; Oklahoma House Bill 197, *Session Laws of Oklahoma 1923*, 296.

23. "Science and Religion," *New York Times*, April 5, 1925, E4; Kenneth K. Bailey, "The Antievolution Crusade of the Nineteen-Twenties" (PhD dissertation, Vanderbilt University, 1954), 67–68.

24. Ibid., 68–69.

25. Summers Amendment, 68th Cong., 1st sess., *Congressional Record* 65 (May 3, 1924): H 7796.

26. West Virginia House Bill 175; *West Virginia Bills of the House of Delegates 1925*, 66.

27. North Carolina House Resolution 10, *Journal of the House of Representatives of the General Assembly of the State of North Carolina 1925*, 18, 203, 225, 280, 290–91.

28. James C. Young, "College Professors, Scientists and Fundamentalists Prepare to Do Battle," *New York Times*, January 30, 1927, XX3; Wilhelm, "Chronology and Analysis," 355.

29. Florida House Bill 691, *Journal of the Florida House of Representatives 1925*, 1267, 1579–80.

30. Texas House Bill 378, *Texas House Journal 1925*, 386; Bailey, "The Antievolution Crusade," 71–72.

31. Tennessee House Bill 185, *House Journal of the Sixty-Fourth General Assembly of the State of Tennessee 1925*, 180, 201, 210, 248, 261, 268, 648, 655, 741; Wilhelm, "Chronology and Analysis," 344–45; Tennessee House Bill 185, *Senate Journal of the Sixty-Fourth General Assembly of the State of Tennessee 1925*, 516–17.

32. Alabama Senate Resolution 55, *Alabama Journal of the Senate 1923*, 1211; Wilhelm, "Chronology and Analysis," 331; House Concurrent Resolution 7, *Journal of the House of Representatives of the State of Florida 1923*, 483.

33. Texas House Concurrent Resolution 6, *Journal of the House of Representatives of the Third Called Session of the Thirty-Eighth Legislature of Texas 1923*, 73–74, 83; Georgia House Resolution 58.246-B, *Journal of the House of Representatives of the State of Georgia 1923*, 353–54; Georgia House Resolution 93.390-C, *Journal of the House of Representatives of the State of Georgia 1923*, 553; Georgia House Resolution 93, *Journal of the House of Representatives of the State of Georgia 1923*, 1001–2.

34. Iowa House File 657, *State of Iowa 1923 Journal of the House*, 758, 1048.

35. Guy H. Fish to Bryan, 30 January 1923, Bryan Papers.

36. Tennessee House Bill 252, *House Journal of the Sixty-Fourth General Assembly of the State of Tennessee 1925*, 225, 233, 318, 423–24, 439.

37. The following bills should be considered to be narrowly antievolution: 1922: South Carolina 1922 Appropriations Bill Amendment; 1923: West Virginia House Bill 153, Georgia House Resolution 58; 1924: Georgia House Bill 731; 1925: North Carolina House Resolution 10, Tennessee Senate Bill 133, West Virginia House Bill 175, Texas House Bill 378, Florida House Bill 691. The following were more broadly worded: 1922: Kentucky House Bill 260, Kentucky House Bill 191, Kentucky Senate Bill 136; 1923: Iowa House File 657, Texas House Bill 97, Oklahoma House Bill 197, Alabama Senate Joint Resolution 55, Tennessee Senate Bill 681, Tennessee House Bill 947, Florida House Conference Resolution 7, Texas House Conference Resolution 6, Georgia House Resolution 93; 1924, US Congress Amendment to Appropriation Bill; 1925: Tennessee House Bill 185, Tennessee House Bill 252, Georgia Amendment. See wording of bills in Wilhelm, "Chronology and Analysis," 314–55. Wilhelm did not include Iowa House File 657; see *State of Iowa 1923 Journal of the House*, 758, 1048.

38. Milton L. Rudnick, *Fundamentalism and the Missouri Synod: a Historical Study of their Interaction and Mutual Influence* (St. Louis, MO: Concordia Publishing, 1966), 75, 79.

39. Timothy Walch, *Parish School: American Catholic Parish Education from Colonial Times to the Present* (New York: Crossroad, 1996), 69–71, 88.

40. Edward Larson, *Summer for the Gods: The Scopes Trial and America's Continuing Debate over Science and Religion* (Cambridge, MA: Harvard University Press, 2001), 126–27, 262.

41. William L. Hornsby, SJ to Bryan, 16 June 1925, Bryan Papers.

42. Callahan quoted in Philip Kinsley, "Truth Goes on Trial Today," *Chicago Daily Tribune*, July 10, 1925, 1.

43. Lavelle in "School Bible Work Urged by Lavelle," *New York Times*, September 9, 1929, 21.

44. Ronald L. Numbers, *The Creationists: The Evolution of Scientific Creationism* (Berkeley, CA: University of California Press, 1993), 55, 72.

45. Alfred Watterson McCann, *God—Or Gorilla: How the Monkey Theory of Evolution Exposes Its Own Methods, Refutes Its Own Principles, Denies Its Own Inferences, Disproves Its Own Case* (New York: Devin-Adair, 1922), 272.

46. Alfred McCann to Bryan, n.d., Bryan papers.

47. Bryan, *The Bible and Its Enemies*, 43.

48. Michael McGerr, *A Fierce Discontent: The Rise and Fall of the Progressive Movement in America, 1870–1920* (New York: Oxford University Press, 2003), 304–5.

49. "Man Was Never a Jelly Fish, Declares Writer in Attack on Evolution Theory," *Imperial Nighthawk* 1 (May 16, 1923): 2.
50. Michael Lienesch, *In the Beginning: Fundamentalism, the Scopes Trial, and the Making of the Antievolution Movement* (Chapel Hill, NC: University of North Carolina Press, 2007), 83–114.
51. T. T. Martin, *Hell and the High School: Christ or Evolution Which?* (Kansas City, MO: Western Baptist Publishing Co., 1923), 10, 62.
52. Martin, *The Evolution Issue*, 46–47.
53. S. J. Betts, untitled pamphlet rebuttal to William Louis Poteat's *Can a Man Be a Christian Today?* (n.p., n.d.) Riley papers, Billy Graham Center Archives [microform].
54. Willard B. Gatewood Jr., *Preachers, Pedagogues and Politicians: The Evolution Controversy in North Carolina, 1920–1927* (Chapel Hill, NC: University of North Carolina Press, 1966), 153.
55. . Gene A. Getz, MBI: *The Story of Moody Bible Institute* (Chicago, IL: Moody Press, 1969), 261–62; William M. Runyan, ed. *Dr. Gray at Moody Bible Institute* (New York: Oxford University Press, 1935), 103–4; Gregg Quiggle, "Moody Magazine" in *Popular Religious Magazines*, ed. P. Mark Fackler and Charles H. Lippy (Westport, CT: Greenwood Press, 1995), 342–47.
56. George McCready Price, "Modern Problems in Science and Religion," *Moody Monthly* 21 (February, March, and April, 1921): 255–57; 303–5; 353–55. For Price's role in the 1920s fundamentalist movement, see Numbers, *The Creationists* (1993), 96–101.
57. Mabel E. Kerr, "Why Not Be Up-to-date?" *Moody Monthly* 23 (April 1923): 336.
58. Bryan, *The Menace of Darwinism*, 7.
59. William R. Straughn (Principal of the State Normal School of Pennsylvania) to Bryan, May 1921, Bryan Papers; A. A. Murphree (president of the University of Florida, Gainesville) to Bryan, 11 February 1922, Bryan Papers.
60. Dennis Joseph Dougherty (Archbishop of Philadelphia) to Bryan, 1 April 1922, Bryan Papers; Florence D. Wills to Bryan, 21 November 1922, Bryan Papers. Bryan also sent hundreds of copies of the pamphlet to prominent fundamentalists such as Charles Trumbull, editor of the *Sunday School Times*, and fundamentalist intellectual W. H. Griffith Thomas. See Charles Trumbull to Bryan, 6 February 1922, Bryan Papers; W. H. Griffith Thomas to Bryan, 17 March 1921, Bryan papers.
61. William Jennings Bryan, "God and Evolution," *New York Times*, February 26, 1922, 1, 11.
62. Guy Viskniskki [of the Republic Syndicate] to Bryan, 5 December 1922, Bryan Papers; *The Commoner* 23 (April 1923): 1.
63. Gatewood, *Preachers, Pedagogues, and Politicians*, 56.
64. LeRoy Johnson, "The Evolution Controversy During the 1920's" (PhD dissertation, New York University, 1954), 103–19.
65. *Journal of the Kentucky Senate 1922*, 330–31.
66. Bailey, "The Anti-Evolution Crusade," 55–57.
67. Kentucky House Bill 191, *Journal of the House of Representatives of the Commonwealth of Kentucky 1922*, 1668–69.

68. Amendment to Kentucky Senate Bill 136, *Journal of the Kentucky Senate 1922*, 1062.
69. George W. Ellis to Bryan and J. W. Porter, 13 March 1922, Bryan Papers.
70. "The Menace of Fundamentalism," *The Independent*, May 30, 1925, 602.
71. G. W. Moothart to Bryan, 5 December 1922, Bryan Papers; J. J. Walters [Governor of Oklahoma] to Bryan, 26 May 1923, Bryan Papers.
72. *Session Laws of Oklahoma 1923*, 292–98.
73. *West Virginia Bills of the House of Delegates, Regular and Extended Sessions*, 1923, 726.
74. John S. Darst to Bryan, 26 January 1923, Bryan Papers; West Virginia resolution, 13 March 1923, Bryan Papers; "Bryan Renews War on Darwin Theory," *New York Times*, April 14, 1923, 6; *Journal of the House of Delegates of the State of West Virginia 1923*, 1048–49.
75. E. L. Simpson [Georgia state legislator] to Bryan, 10 July 1923, Bryan Papers; Maynard Shipley, *The War on Modern Science* (New York: Alfred A. Knopf, 1927), 171; Wilhelm, "Chronology and Analysis," 324–25; Guy H. Fish [of the Des Moines Bible Association on Christian Fundamentals] to Bryan, 30 January 1923, Bryan Papers; Florida Senate Concurrent Resolution 7, *Journal of the House of Representatives of the State of Florida 1923*, 1789–90, 1840, 2052; J. T. Stroder [Texas state legislator] to Bryan, 18 June 1923, Bryan Papers; Texas House Concurrent Resolution 4, *Journal of the House of Representatives of the Second Called Session of the Thirty-Eighth Legislature of Texas 1923*, 139.
76. Guy H. Fish to Bryan, 30 January 1923, Bryan Papers.
77. J. Frank Norris, "Address to Texas State Legislature," *The Searchlight* 6 (February 23, 1923): 1.
78. Norris quoted by James Gray, "Will the Christian Taxpayers Stand for This?" *Moody Monthly* 23 (May 1923): 409.
79. J. Frank Norris, "Report from the Fifth World's Fundamental Convention," *CFSC* 5 (July-September 1923): 4.
80. William V. Trollinger, *God's Empire: William Bell Riley and Midwestern Fundamentalism* (Madison, WI: University of Wisconsin Press, 1990), 48; Pat M. Neff to Bryan, 16 March 1923, Bryan papers; J. T. Stroder to Bryan, 24 June 1923, Bryan Papers.
81. Bryan to W. J. Singleterry [of the Florida State Senate], 11 April 1923, Bryan Papers; W. J. Singleterry to Bryan, 12 April 1923, Bryan Papers.
82. Florence D. Wills to Bryan, 21 November 1922, Bryan Papers.
83. Florida House Concurrent Resolution 7, *Journal of the House of Representatives of the State of Florida 1923*, 483.
84. Bailey, "The Antievolution Crusade," 77.
85. Trollinger, *God's Empire*, 48.
86. Bailey, "The Antievolution Crusade," 81–82; Johnson, "The Evolution Controversy," 184.
87. For more about Tennessee's long struggle over the issue of evolution and schools, see Charles A. Israel, *Before Scopes: Evangelicalism, Education, and Evolution in Tennessee, 1870–1925* (Athens, GA: University of Georgia Press, 2004); Tennessee House Bill 185, *House Journal of the Sixty-Fourth General Assembly of the State of Tennessee 1925*, 180, 201, 210, 248, 261, 268, 648, 655, 741; Tennessee House

Bill 185, *Senate Journal of the Sixty-Fourth General Assembly of the State of Tennessee 1925*, 221, 225, 251, 254, 286, 466, 477, 516–17; Tennessee Senate Bill 133, *Senate Journal of the Sixty-Fourth General Assembly of the State of Tennessee 1925*, 121, 136, 214, 252, 466; Wilhelm, "Chronology and Analysis," 344.

88. John A. Shelton to Bryan, 5 February 1925, Bryan Papers; Bryan to Shelton, 9 February 1925, Bryan Papers.
89. "Paint W. J. Bryan as a 'Medievalist,'" *New York Times*, March 2, 1922, 12. Many other writers similarly accused Bryan of "medievalism." See Harry Emerson Fosdick, "Mr. Bryan and Evolution," *New York Times*, March 12, 1922; Frederick F. Shannon, "Bryanism," *Christian Century* 39 (March 23, 1922): 428–31.
90. "Shaw Calls Ideas of Bryan 'Infantilism,'" *New York Times*, June 10, 1925, 1.
91. "Heresy-Hunting Spreads in the South," *Christian Century* 39 (January 26, 1922).
92. "Reactionaries Fight Education," *Christian Century* 39 (February 23, 1922).
93. Arthur Wilford Nagler, "Fundamentalism in History," *The Methodist Review* (September 1923): 673.
94. "The Menace of Fundamentalism," *The Independent*, May 30, 1925, 602.

Chapter 5

1. Clarence Darrow, William J. Bryan, et al., *The World's Most Famous Court Trial: Tennessee Evolution Case* (Cincinnati, OH: National Book Company, 1925), 282.
2. The best source for understanding this trial is Edward Larson, *Summer for the Gods: The Scopes Trial and America's Continuing Debate over Science and Religion* (Cambridge, MA: Harvard University Press, 2001). Another good source is Jeffrey P. Moran, *The Scopes Trial: A Brief History with Documents* (Boston: Bedford/St. Martin's Press, 2002). Also helpful is Douglas O. Linder, "Famous Trials: Tennessee vs. John Scopes: 'The Monkey Trial,'" University of Missouri, Kansas City School of Law, http://www.law.umkc.edu/faculty/projects/ftrials/scopes.htm (accessed July 18, 2005).
3. Larson, *Summer for the Gods*, 171, 174; *Tennessee House Journal 1925*, 225, 233, 318, 423–25, 439.
4. Moran, *Scopes Trial*, 156.
5. William Jennings Bryan, "Who Shall Control?" typescript statement in Bryan Papers, file 3, Dayton Trial Correspondence, June 24 to June 30, 1925, Library of Congress, Washington, DC.
6. Governor Austin Peay, "Statement," *Moody Bible Institute Monthly* [*Moody Monthly*] 25 (June 1925): 462.
7. James M. Gray to Bryan, 22 June 1925, Bryan Papers.
8. Larson, *Summer for the Gods*, 156.
9. Harry Emerson Fosdick, "Shall the Fundamentalists Win?" *Christian Century* 39 (June 8, 1922): 716.
10. F. Z. Brown, "The False Premise of Dr. Fosdick's Farewell Sermon," *Moody Monthly* 25 (June 1925): 453.
11. Harry Boehme, "Can the Fundamentalists Lose?" *Christian Fundamentals in School and Church* [*CFSC*] 5 (October–December 1922): 29.
12. Larson, *Summer for the Gods*, 203.

13. Ibid., 142.
14. Moran, *Scopes Trial*, 28.
15. "Cranks and Freaks Flock to Dayton," *New York Times*, July 11, 1925, 1–2.
16. Charles McD. Puckette, "The Evolution Arena at Dayton," *New York Times*, July 5, 1925, SM1, 22.
17. "Evolution Trial Raises Two Sharp Issues," *New York Times*, May 31, 1925, XX4.
18. Larson, *Summer for the Gods*, 93.
19. "Cranks and Freaks Flock to Dayton," 2.
20. "Fights Evolution to Uphold Bible," *New York Times*, July 5, 1925, 4.
21. "Farmers Will Try Teacher," *New York Times*, July 11, 1925, 1.
22. "Cranks and Freaks Flock to Dayton," 1, 2.
23. H. L. Mencken, "The Hills of Zion," in *Prejudices: Fifth Series* (New York: Octagon Books, 1977), 82.
24. Henry M. Hyde, "Jury for Scopes Trial Selected; Recess Is Taken," *Baltimore Sun*, July 11, 1925, 1, in *Scopes Trial*, ed. Moran, 78.
25. Russell Owen, "Dayton's Remote Mountaineers Fear Science," *New York Times*, July 19, 1925, XX3.
26. Henry D. Shapiro, *Appalachia on Our Mind: The Southern Mountains and Mountaineers in the American Consciousness, 1870–1920* (Chapel Hill, NC: University of North Carolina Press, 1978), 5, 17, 99, 115, 240.
27. Maynard Shipley, *The War on Modern Science* (New York: Alfred A. Knopf, 1927), 357.
28. T. S. Stribling, *Teeftallow*, in *Controversy in the 'Twenties: Fundamentalism, Modernism, and Evolution*, ed. Willard B. Gatewood Jr. (Nashville, TN: Vanderbilt University Press, 1969), 376.
29. "Decision Today is Likely," *New York Times*, July 14, 1925, 1.
30. H. L. Mencken, "Battle Now Over, Mencken Sees; Genesis Triumphant and Ready for New Jousts," *Baltimore Evening Sun*, July 18, 1925, 1, in *Scopes Trial*, ed. Moran, 170.
31. Kenneth K. Bailey, "The Antievolution Crusade of the Nineteen-Twenties" (PhD dissertation, Vanderbilt University, 1954), 242.
32. "An Anti-Evolution Bill," *New York Times*, January 30, 1927, E5.
33. "Farmers Will Try Teacher," 1.
34. "Decision Today is Likely" *New York Times*, July 14, 1925, 1.
35. Moran, *Scopes Trial*, 92.
36. Moran, *Scopes Trial*, 35–39. See also Donald E. Boles, *The Bible, Religion, and the Public Schools* (Ames, IA: Iowa State University Press, 1963).
37. Moran, *Scopes Trial*, 39.
38. Ibid., 40–43.
39. Larson, *Summer for the Gods*, 130–35.
40. Ibid., 185–87.
41. Moran, *Scopes Trial*, 47.
42. Larson, *Summer for the Gods*, 187–89. Hays is quoted on page 189.
43. Linder, "Famous Trials: Tennessee vs. John Scopes: 'The Monkey Trial.'"
44. Ibid.
45. Moran, *Scopes Trial*, 154.

46. Larson, *Summer for the Gods*, 191–93.
47. J. Frank Norris, "Bryan Wins Greatest Victory of his Career—Bible Triumphs Over Infidelity: Commoner Outwits Darrow in Dayton Evolution Trial," *The Searchlight* 8 (July 24, 1925): 1.
48. "Big Crowd Watches Trial Under Trees," *New York Times*, July 21, 1925, 1.
49. Larson, *Summer for the Gods*, 201.
50. Owen, "Dayton's Remote Mountaineers Fear Science," XX3.
51. "The Scopes Trial," *The Independent*, July 25, 1925, 115, 3921.
52. Frederick Lewis Allen, *Only Yesterday: An Informal History of the Nineteen-Twenties* (New York: Harper and Bros., 1931), 206.
53. Duncan Aikman, "Ape Laws as Political Medicine," *The Independent*, May 8, 1926, 116.
54. Shipley, *War on Modern Science*, 187.
55. Ibid., 3–4.
56. Maynard Shipley, "Evolution Still a Live Issue In the Schools," *Current History* 27 (March 1928): 801.
57. H. L. Mencken, "In Memoriam: WJB," in *Prejudices, Fifth Series* (New York: Octagon Books, 1977), 68.
58. Ibid., 71.
59. H. L. Mencken, "Jacquerie," in *The Bathtub Hoax and Other Blasts and Bravos* (New York: Alfred A. Knopf, 1958), 136–40.
60. Curtis Lee Laws, "Editorial Notes and Comments," *Watchman-Examiner* 13 (August 20, 1925): 1071.
61. William Bell Riley, "William Jennings Bryan University," *CFSC* 7 (October–December 1925): 52.
62. T. C. Horton, "Bryan's Benediction," *King's Business* (December 1925): 534–35; T. C. Horton, "Bryan the Brave—'Defender of the Faith,'" *King's Business* 16 (September 1925): 372.
63. Philip E. Howard, "William Jennings Bryan as his Friends Knew Him," *Sunday School Times* 67 (August 8, 1925): 499.
64. Alabama House Bill 30, *Journal of the House of Representatives of the State of Alabama 1927*, 89; Alabama House Bill 969, *Journal of the House of Representatives of the State of Alabama 1927*, 1622, 1983, 2040; Alabama House Bill 1103, *Journal of the House of Representatives of the State of Alabama 1927*, 2153, 2426–27, 2596; Arkansas House Bill 34, *Journal of the House of Representatives for the General Assembly of the State of Arkansas 1927*, 68–69, 263, 323–24; California Assembly Bill 145, *Journal of the Assembly during the Forty-Seventh Session of the Legislature of the State of California 1927*, 182, 445, 482, 543–44, 566, 2 639–41; New Hampshire House Bill 268, "Journal of the House of Representatives 1927," *Journals New Hampshire Senate and House 1927*, 154, 274; North Dakota House Bill 222, *State of North Dakota Journal of the House of the Twentieth Session of the Legislative Assembly 1927*, 519, 1022; Oklahoma House Bill 81, *Journal of the House of Representatives of the Legislature of the State of Oklahoma 1927*, 281, 305; South Carolina House Bill 60, *Journal of the House of Representatives of the First Session of the 78th General Assembly of the State of South Carolina 1927*, 70, 1128, 1482; West Virginia House Resolution, *Journal of the House of Delegates of West Virginia*

1927, 65, 97–98; West Virginia House Bill 264; *Journal of the House of Delegates of West Virginia 1927*, 104, 663; West Virginia House Bill 358, *Journal of the House of Delegates of West Virginia 1927*, 129; North Carolina House Bill 263, *Journal of the House of Representatives of the General Assembly of the State of North Carolina 1927*, 85, 241; Delaware House Bill 92, *Delaware House Journal 1927*, 156; Maine House Paper 834, *Legislative Record of the Eighty-Third Legislature of the State of Maine 1927*, 239, 242, 247–49, 313, 835; Minnesota Senate Bill 701, *Journal of the Senate of the State of Minnesota 1927*, 508–9; Florida House Bill 87, *Florida House Journal 1927*, 3000–3001; Richard David Wilhelm, "A Chronology and Analysis of Regulatory Actions Relating to the Teaching of Evolution in Public Schools" (PhD dissertation, University of Texas–Austin, 1978), 373 [Missouri House Bill No. 89].

65. See Jonathan Zimmerman, *Distilling Democracy: Alcohol Education in America's Public Schools, 1880–1925* (Lawrence, KS: University Press of Kansas, 1999).

66. See Stewart Cole, *The History of Fundamentalism* (Westport, CT: Greenwood Press, 1971); Norman F. Furniss, *The Fundamentalist Controversy, 1918–1931* (New Haven, CT: Yale University Press, 1954); Richard Hofstadter, *Anti-Intellectualism in American Life* (New York: Alfred A. Knopf, 1963); H. Richard Niebuhr, "Fundamentalism," in *The Encyclopedia of the Social Sciences*, ed. Edwin R. A. Seligman (New York: Macmillan Co., 1931), 526–27.

67. George M. Marsden, *Fundamentalism and American Culture*, 7, 15, 18–21, 24, 26, 57–62, 65, 112, 120–21, 126, 169, 174, 209, 212–21, 226–27.

68. William Bell Riley and Charles Smith, "Should Evolution Be Taught in Tax Supported Schools?: Riley-Smith Debate" (n.p., n.d.), 1.

69. T. T. Martin, *Hell and the High School: Christ or Evolution, Which?* (Kansas City, MO: Western Baptist Publishing Co., 1923), 72. Emphasis in original.

70. Ronald L. Numbers, "Reading the Book of Nature through American Lenses," in *Science and Christianity in Pulpit and Pew* (New York: Oxford University Press, 2007), 62–68.

71. Alfred Fairhurst, *Atheism in Our Universities* (Cincinnati, OH: Standard Publishing Co., 1923), 71.

72. George Wilson McPherson, *The Crisis in Church and College* (Yonkers, NY: Yonkers Book Co., 1919), 119.

73. Quoted in Virginia L. Brereton, *Training God's Army: The American Bible School, 1880–1940* (Bloomington, IN: Indiana University Press, 1990), 179n19.

74. William Jennings Bryan, *In His Image* (New York: Fleming H. Revell Co., 1922), 93, 94.

75. James Gilbert, *Redeeming Culture: American Religion in an Age of Science* (Chicago, IL: University of Chicago Press, 1997), 24–27; Burton E. Livingston [permanent secretary to the American Association for the Advancement of Science] to William Jennings Bryan, 29 September 1924, Bryan Papers, file 1924.

76. William Jennings Bryan, "God and Evolution," *New York Times*, February 26, 1922, 1.

77. Ronald L. Numbers, *Darwinism Comes to America* (Cambridge, MA: Harvard University Press, 1998), 33.

78. Sue Hicks to Ira Hicks, quoted in Larson, *Summer for the Gods*, 130.

79. Bryan to Sue Hicks, quoted in Gilbert, *Redeeming Culture*, 32.

80. Bryan to Howard A. Kelly, 10 June 1925, Bryan Papers.

81. J. Frank Norris to Bryan, June or July 1925, Bryan Papers, file 1925.

82. George McCready Price, "Modern Scientific Discoveries," *CFSC* 5 (October–December 1922): 74.

83. George McCready Price, "Modern Problems in Science and Religion," *Moody Monthly* 21 (February 1921): 256.

84. James M. Gray, "Editorial Notes," *Moody Monthly* 21 (November 1920): 101.

85. G. M. Price to Bryan, 27 February 1922, Bryan Papers.

86. Leander S. Keyser, "Seeking for Obscure Causes," *Moody Monthly* 22 (August 1922): 1139.

87. Moran, *Scopes Trial*, 40–41.

88. T. T. Martin, *The Evolution Issue* (Los Angeles, CA: n.p., 1923?), 19; Martin, *Hell and the High School*.

89. Fairhurst, *Atheism in Our Universities*, 42.

90. Riley and Smith, "Should Evolution Be Taught in Tax Supported Schools?" 2.

91. Ronald L. Numbers, *The Creationists: The Evolution of Scientific Creationism* (Berkeley, CA: University of California Press, 1993), 54–55.

92. Dennis Royal Davis, "Presbyterian Attitudes Toward Science and the Coming of Darwinism in America, 1859 to 1929" (PhD dissertation, University of Illinois at Urbana–Champaign, 1980), 125–26; Jon H. Roberts, *Darwinism and the Divine: Protestant Intellectuals and Organic Evolution, 1859–1900* (Madison, WI: University of Wisconsin Press, 1988), 125, 138.

93. Arthur I. Brown, "Evolution and the Bible" (1922), in *The Antievolution Works of Arthur I. Brown*, ed. Ronald L. Numbers (New York: Garland Publishing, Inc., 1995), 10.

94. Numbers, *The Creationists* (2006), 22–23.

95. Ibid., 19.

96. Bryan to C. B. McMullen, 7 June 1925; Bryan to H. A. Kelly, 10 June 1925; H. A. Kelly to Bryan, 15 June 1925; Bryan to Kelly, 22 June 1925; Alfred McCann to Bryan, n.d.; Bryan to Kelly, 17 June 1925; Bryan Papers. See also Numbers, *The Creationists* (2006), 88–89; Larson, *Summer for the Gods*, 129–31.

97. Bryan to G. M. Price, 7 June 1925; G. M. Price to Bryan, 1 July 1925, Bryan Papers. See also Numbers, *Creationists* (2006), 116–19; Larson, *Summer*, 130.

98. Moran, *Scopes Trial*, 49.

99. Numbers, *Creationists* (2006), 91.

100. Ibid., 111.

101. Numbers, *Darwinism Comes to America*, 24–48.

Chapter 6

1. Quoted in Edward Larson, *Summer for the Gods: The Scopes Trial and America's Continuing Debate over Science and Religion* (Cambridge, MA: Harvard University Press, 2001), 215.

2. Ibid., 218.

3. 69th Cong., 1st sess., *Congressional Record* 67 (March 16, 1926): H 5748.

4. "Evolution Battle to Go to Congress; New Law is Sought," *New York Times*, July 24, 1925, 1.

5. 69th Cong., 1st sess., *Congressional Record* 67 (March 16, 1926): H 5747–48.

6. Ibid., 5749.

7. James C. Young, "College Professors, Scientists and Fundamentalists Prepare to Do Battle," *New York Times*, January 30, 1927, XX3.

8. Kentucky House Bill 96, *Journal of the House of Representatives of the Commonwealth of Kentucky 1926*, 173.

9. Louisiana House Bills 41, 208, 279, 314, *Official Journal of the Proceedings of the House of Representatives of the State of Louisiana at the Third Regular Session of the Legislature 1926*, 11, 34, 99–100, 130, 167, 184, 190, 219, 574–75, 642–43, 675, 700, 762, 782, 833; Maynard Shipley, *The War on Modern Science* (New York: Alfred A. Knopf, 1927), 79.

10. Kenneth K. Bailey, "The Antievolution Crusade of the Nineteen-Twenties" (PhD dissertation, Vanderbilt University, 1954), 224.

11. Larson, *Summer for the Gods*, 231.

12. Bailey, "The Antievolution Crusade," 250. See also Gerald Skoog, "Topic of Evolution in Secondary School Biology Textbooks: 1900–1977," *Science Education* 63 (1979): 621–40.

13. C. H. Thurber to Bryan, 21 November 1923, Bryan Papers, Library of Congress, Washington, DC; Bryan to Thurber, 22 December 1923, Bryan Papers.

14. Richard David Wilhelm, "A Chronology and Analysis of Regulatory Actions Relating to the Teaching of Evolution in Public Schools" (PhD dissertation, University of Texas–Austin, 1978), 344–45.

15. Bailey, "Antievolution Crusade," 224–30, 247; Maynard Shipley, "A Year of the Monkey War," *The Independent*, October 1, 1927, 326–28, 344–45; Young, "College Professors, Scientists and Fundamentalists Prepare to Do Battle," XX3; "Protest in Florida on Anti-Darwin Bill," *New York Times*, April 30, 1927, 3; "Missouri Kills Anti-Evolution Bill," *New York Times*, February 9, 1927, 8; "Act on Evolution Bills," *New York Times*, February 10, 1927, 38; "Beat Anti-Evolution Bill," *New York Times*, April 14, 1927, 24.

16. William Bell Riley, "Five Addresses on Evolution in the State of Minnesota," *Christian Fundamentals in School and Church* [*CFSC*] 9 (January-March 1927): 20–21.

17. "Professors to War on Evolution Laws," *New York Times*, January 1, 1927, 1.

18. *Legislative Record of the Eighty-Third Legislature of the State of Maine 1927*, February 18, 1927, 248.

19. Ibid., 249.

20. Alabama House Bill 30, *Journal of the House of Representatives of the State of Alabama 1927*, 89; Wilhelm, "Chronology and Analysis," 368.

21. South Carolina House Bill 60, in Wilhelm, "Chronology and Analysis," 374; Arkansas House Bill 34, *Journal of the House of Representatives for the General Assembly of the State of Arkansas 1927*, 68–69, 263, 323–24.

22. North Dakota House Bill 222, *State of North Dakota Journal of the House of the Twentieth Session of the Legislative Assembly 1927*, 519, 1022.

23. West Virginia House Bill 264; *Journal of the House of Delegates of West Virginia 1927*, 104; Florida House Bill 87, *Florida House Journal 1927*, 3000–3001; *Journal of the State Senate of Florida of the Session of 1927*, 2411–12.

24. The following bills should be considered to be narrowly antievolution: 1926: Kentucky House Bill 96, Louisiana House Bill 41, Louisiana House Bill 279, Mississippi Senate Bill 214, Mississippi House Bill 39; 1927: California Assembly Bill 145, Missouri House Bill 89, West Virginia Resolution, Delaware House Bill 92, Maine House Paper 834, Minnesota Senate Bill 701, Minnesota House Bill 837, Alabama House Bill 969, Alabama House Bill 1103, West Virginia House Bill 358; 1928: Arkansas Act 1; 1929: Texas House Bill 90. The following were more broad: 1926: Mississippi House Bill 77, US Congress Amendment to Appropriation Bill, Louisiana House Bill 208, Louisiana House Bill 314; 1927: Alabama House Bill 30, Arkansas House Bill 34, Oklahoma House Bill 81, South Carolina House Bill 60, West Virginia House Bill 264, New Hampshire House Bill 280, North Carolina House Bill 263, North Dakota House Bill 227, Florida House Bill 87. See text of most bills in Wilhelm, "Chronology and Analysis," 356–97. Wilhelm did not include two Mississippi bills, Senate Bill 214 (1926) and House Bill 39 (1926).

25. Alabama House Bill 30, *Journal of the House of Representatives of the State of Alabama 1927*, 89.

26. U.S. GenWeb Archives, "Pickens County Al Archives Cemeteries, Holly Springs Cemetery Survey," http://files.usgwarchives.org/al/pickens/cemeteries/hollyspr420gcm .txt (accessed March 13, 2009).

27. Alabama House Bill 969, *Journal of the House of Representatives of the State of Alabama 1927*, 1622, 1983, 2040; Alabama House Bill 1103, *Journal of the House of Representatives of the State of Alabama 1927*, 2153, 2426–27, 2596.

28. Arkansas House Bill 34, *Journal of the House of Representatives for the General Assembly of the State of Arkansas 1927*, 68–69, 263, 323–24; Bailey, "Antievolution Crusade," 243–47; Virginia Gray, "Anti-Evolution Sentiment and Behavior: The Case of Arkansas," *Journal of American History* 62 (1970): 353–65; Shipley, "Year of the Monkey War," 326.

29. William Bell Riley, "The Prospect of a Great Fundamentalist University," *CFSC* 8 (January-March 1926): 24–25.

30. *King's Business* (November 1920): 1025.

31. George M. Marsden, *Fundamentalism and American Culture, The Shaping of Twentieth-Century Evangelicalism, 1870–1925* (New York: Oxford University Press, 1980), 189, 191.

32. Joel A. Carpenter, *Revive Us Again*: *The Reawakening of American Fundamentalism* (New York: Oxford University Press, 1997), 102.

33. Washburn to Bryan, 13 July, 1925; 18 July, 1925, Bryan Papers.

34. "National Body Selects Clearwater as Head Washburn President," *CFSC* 8 (January–March 1926): 30.

35. Washburn to Bryan, 24 July, 1925, Bryan Papers.

36. T. T. Martin, *King's Business* 17 (January–June 1926): 77.

37. Shipley, "Year of the Monkey War," 326.

38. "The Bible Crusader's [*sic*] Challenge" *CFSC* 8 (April–June 1926): 56.

39. "Editorial," *King's Business*, 17 (May 1926): 297.

40. George Washburn, *Moody Bible Institute Monthly* [*Moody Monthly*] 26 (January 1926): 229.

41. "The Bible Crusader's Challenge," 53–56.

42. Shipley, *War on Modern Science*, 28.

43. George Washburn to Mrs. John Roach Straton, 13 May 1929, John Roach Straton Papers, American Baptist—Samuel Colgate Historical Library; American Baptist Historical Society, Rochester, NY.

44. 'Defenders Editorial Staff,' *Fire by Night and Cloud By Day: A History of the Defenders of the Christian Faith* (Wichita, KS: Defenders, Inc., 1966), 9.

45. Leo P. Ribuffo, *The Old Christian Right: The Protestant Far Right from the Great Depression to the Cold War* (Philadelphia: Temple University Press, 1983), 89; "The Defenders," *CFSC* 8 (April–June 1926): 8; Gerald B. Winrod, "The Fight in Kansas," *CFSC* 8 (October–December 1926): 11; *The Christian Fundamentalist* 1 (November 1927): 22; *Fire by Night*, 5–24.

46. William Bell Riley, "Mr. Bryan and the Scopes Trial," *CFSC* 9 (January–March 1927): 9.

47. Ribuffo, *Old Christian Right*, 90

48. H. L. Mencken, *Prejudices: Fifth Series* (New York: Octagon Books, 1977), 111.

49. *Fire By Night*, 8.

50. Ibid., 10–11.

51. William G. Shepherd, "Ku Klux Koin," *Collier's* (July 21, 1928): 8–9, 38–39.

52. Harbor Allen, "'Supreme Kingdom's Campaign," *CFSC* 8 (October–December 1926): 51.

53. "Dr. Straton on His Supreme Kingdom Connections," *New York Herald-Tribune*, January 20, 1927; "Dr. Straton's Statement Regarding His Connection with the Supreme Kingdom," delivered to Calvary Baptist and the New York Baptist Bible Union, January 19, 1927, typescript copy in Straton Papers; Straton to W. T. Anderson, 8 Feb 1927, Straton Papers.

54. Shipley, *The War on Modern Science*, 48.

55. Ibid., 44.

56. Riley, "Five Addresses on Evolution in the State University of Minnesota," 20–21.

57. Wilhelm, "Chronology and Analysis," 356–57; Mississippi Senate Bill 214, *Journal of the Senate of the State of Mississippi 1926*, 436, 646; Mississippi House Bill 39, *Journal of the Senate of the State of Mississippi 1926*, 294, 589, 662–66, 677, 1009, 1017, 1303, 1308, 1630, 1676, 1680–81; Mississippi House Bill 77, *Journal of the Senate of the State of Mississippi 1926*, 634–35, 703, 723–25, 810, 919–21, 943, 975, 1617.

58. "Editorial," *King's Business* 17 (May 1926): 297; James M. Gray, "Editorial Notes," *Moody Monthly* 26 (May 1926).

59. Wilhelm, "Chronology and Analysis," 356.

60. "Editorial," *King's Business* 17 (May 1926): 297.

61. James M. Gray, "Editorial Notes," *Moody Monthly* 26 (May 1926).

62. Bailey, "The Antievolution Crusade," 228.

63. Washburn to Bryan, 24 July, 1925, Bryan Papers; T. T. Martin, *King's Business* 17 (January–June 1926): 77.

64. LeRoy Johnson, "The Evolution Controversy During the 1920's" (PhD dissertation, New York University, 1954), 186–87; Bailey, "The Antievolution Crusade," 224–29.

65. Shipley, *The War on Modern Science*, 65.

66. Wilhelm, "Chronology and Analysis," 356.

67. Willard B. Gatewood Jr., *Preachers, Pedagogues and Politicians: The Evolution Controversy in North Carolina, 1920–1927* (Chapel Hill, NC: University of North Carolina Press, 1966), 30, 40–44, 49–50, 98–101; North Carolina House Resolution 10, *Journal of the House of Representatives of the General Assembly of the State of North Carolina 1925*, 18, 203, 225, 280, 290–91.

68. Gatewood, *Preachers, Pedagogues and Politicians*, 190–94;

69. Johnson, "The Evolution Controversy," 190.

70. Ronald L. Numbers, *The Creationists: From Scientific Creationism to Intelligent Design* (Cambridge, MA: Harvard University Press, 2006), 75.

71. Gatewood, *Preachers, Pedagogues and Politicians*, 197.

72. Ibid., 201.

73. Ibid., 186.

74. Ibid., 191.

75. Ibid., 197.

76. Bess Davenport, "Scorching Epithets Bring Discord into Anti-Evolution War," *Raleigh News and Observer*, May 5, 1926, 1.

77. Ibid., 1, 2, 15.

78. Bailey, "The Antievolution Crusade," 232.

79. Ferenc M. Szasz, "William B. Riley and the Fight Against Teaching of Evolution in Minnesota" *Minnesota History* (Spring 1969): 202, 204, 207, 208; "Pastors Condemn Evolution Theory in Public Schools," *Minneapolis Journal*, October 26, 1922; "The Anti-Evolution League," *CFSC* 5 (January-March 1923): 16; "Minnesota Anti-Evolution League," *CFSC* 5 (April–June 1923): 31–32.

80. William V. Trollinger, *God's Empire: William Bell Riley and Midwestern Fundamentalism* (Madison, WI: University of Wisconsin Press, 1990), 51; Szasz, "William B. Riley and the Fight Against Teaching of Evolution in Minnesota," 212; "The Anti-Evolution Fight in Minnesota," *CFSC* 9 (April–June 1927): 12; "Anti-Evolution Bill Bars Man's Origin Theory," *Minneapolis Tribune*, January 8, 1927; "Riley Upheld in Evolution Debate," *Minneapolis Journal*, February 1, 1927.

81. "Booklets Supporting Anti-Evolution Bill Placed in Circulation," *Minneapolis Tribune*, January 15, 1927.

82. *Why Pass a Law Against Evolution in Minnesota?* (Minneapolis: World's Christian Fundamentals Association, 1927), Riley Papers, Billy Graham Center Archives [microform].

83. Minnesota Senate Bill 701, *Journal of the Senate of the State of Minnesota 1927*, 508–9.

84. "The Anti-Evolution Fight in Minnesota," *CFSC* 9 (April–June 1927): 13.

85. Szasz, "William B. Riley and the Fight Against Teaching of Evolution in Minnesota," 212.

86. Ibid., 213.

87. *Why Pass a Law Against Evolution in Minnesota?* 13.

88. Sydney E. Ahlstrom, *A Religious History of the American People* (New Haven, CT: Yale University Press, 1974), 756–61.

89. Szasz, "William B. Riley and the Fight Against Teaching of Evolution in Minnesota," 215–16.

90. Gray, "Anti-Evolution Sentiment and Behavior: The Case of Arkansas," 355–57.

91. William Bell Riley and Charles Smith, *Should Evolution Be Taught in Tax Supported Schools?* (n.p., n.d.), 22.
92. Bailey, "The Antievolution Crusade," 245.
93. Riley and Smith, *Should Evolution Be Taught in Tax Supported Schools?* 28.
94. Michael Lienesch, *In the Beginning: Fundamentalism, the Scopes Trial, and the Making of the Antievolution Movement* (Chapel Hill, NC: University of North Carolina Press, 2007), 115–38. Tarrow's work is paraphrased on page 116.
95. Ibid., 134–36.

Chapter 7

1. L. L. Bernard, "The Development of the Concept of Progress," *The Journal of Social Forces* 3 (January, 1925): 208.
2. Harry Elmer Barnes, "Sociology and Ethics: A Genetic View of the Theory of Conduct," *The Journal of Social Forces* 3 (January 1925): 214.
3. Willard B. Gatewood Jr., *Preachers, Pedagogues and Politicians: The Evolution Controversy in North Carolina, 1920–1927* (Chapel Hill, NC: University of North Carolina Press, 1966), 114–19.
4. Ibid., 120–22.
5. Ibid., 129–30.
6. Ibid., 122, 159–60.
7. Maynard Shipley, *The War on Modern Science* (New York: Alfred A. Knopf, 1927), 108.
8. T. T. Martin, "A Fight to the Finish Against Evolution in Baylor University," *The Searchlight* (November 10, 1922): 1; J. Frank Norris, "Wake Forest College Making Infidels Out of the Sons of Baptist Ministers," *Searchlight* (July 6, 1923): 2.
9. Gatewood, *Preachers, Pedagogues, and Politicians*, 96.
10. Ibid., 172–75.
11. J. Frank Norris, "Poteat's Son Attacks Bryan, Straton and Norris," *Searchlight* (May 2, 1924): 1.
12. William Louis Poteat, *Can a Man Be a Christian To-day?* (Chapel Hill, NC: University of North Carolina Press, 1925), 14.
13. Ibid., 35.
14. Ibid., 21.
15. Gatewood, *Preachers, Pedagogues, and Politicians*, 175.
16. See scrapbooks in William Bell Riley Papers, Billy Graham Center Archives [microform].
17. Ferenc M. Szasz, "William B. Riley and the Fight Against Teaching of Evolution in Minnesota," *Minnesota History* (Spring 1969): 208. See also William V. Trollinger, *God's Empire: William Bell Riley and Midwestern Fundamentalism* (Madison: University of Wisconsin Press, 1990), 50.
18. "Riley Assails 'U' as Fostering State Atheism," *Minneapolis Tribune*, repr. in *Christian Fundamentals in School and Church* [*CFSC*] (April–June 1926): 33.
19. Ibid., 36.
20. Ibid., 35; Szasz, "William B. Riley and the Fight Against Teaching of Evolution in Minnesota," 209.

21. William Bell Riley, "Five Addresses on Evolution in the State University of Minnesota," *CFSC* (January–March 1927): 20–21. See also Szasz, "William B. Riley and the Fight Against Teaching of Evolution in Minnesota," 210.

22. Ibid., 21.

23. Szasz, "William B. Riley and the Fight Against Teaching of Evolution in Minnesota," 210.

24. William Bell Riley, "The Case Against Carleton," *Christian Fundamentalist* (October 1927): 5.

25. Trollinger, *God's Empire*, 137.

26. William Bell Riley, "Carleton College Divorced by Minnesota Baptist State Convention," *Christian Fundamentalist* (November 1928): 12.

27. Mark A. Greene, "The Baptist Fundamentalists' Case Against Carleton, 1926–1928," *Minnesota History* (Spring 1990): 21–24.

28. L. M. Aldridge, "Arkansas Baptists Oust College President Who Refused to Sign Fundamentalist Confession of Faith," *Searchlight* (November 27, 1925): 1.

29. W. Carl Richards, *Conservative Association of Wooster College Alumni, Students and Friends* (n.p., n.d.), Riley papers.

30. George H. Young, *Alarming Conditions in Kalamazoo College! Modernism and its Consequences! Can Michigan Baptists Continue to Support Kalamazoo?* (n.p., June 24, 1927), Riley papers.

31. C. Allyn Russell, "Thomas Todhunter Shields, Canadian Fundamentalist," *Ontario History* (December 1978): 270. See also T. T. Shields, "How the Lord Visited Toronto," *Moody Bible Institute Monthly* [Moody Monthly] (June 1921): 430–32; "Marshall of McMaster," *Searchlight* (January 29, 1926): 1; William Bell Riley, "The Canadian Conflict," *CFSC* 9 (January–March, 1927): 15–19.

32. *League of Evangelical Students Constitution*, President's Office Papers, Accession 2006-18, L. S. Chafer correspondence, Box 13, File 16, Archives, Dallas Theological Seminary.

33. "Address under the Auspices of the League of Evangelical Students," 21 November 1925, typescript document, J. Gresham Machen papers, Westminster Theological Seminary archives, Philadelphia.

34. "Editorial," *The Evangelical Student* 4 (January 1930): 1.

35. "Editorial," *The Evangelical Student* 4 (April 1930): 1–5 [YMCA]; "Letter to League of Evangelical Students Board of Trustees," n.d., President's Office Papers, Accession 2006-18, L. S. Chafer correspondence, Box 13, File 16, Archives, Dallas Theological Seminary [YMCA]; William H. Hockman, "The Missionary Challenge of the League," *The Evangelical Student* 4 (April 1930): 14 [Student Volunteer Movement].

36. Winfield Burggraaff, "Whither Students?" *The Evangelical Student* 3 (April 1929): 24.

37. *The League of Evangelical Students: Origin, Platform, Organization, Functions, Service* (Princeton, NJ: League of Evangelical Students, 1927), Machen papers.

38. George Johnson, "The League and Evangelism," *The Evangelical Student* 5 (October 1930): 19.

39. A. A. MacRae, "Why the League?" *The Evangelical Student* 1 (April 1926): 3.

40. George Johnson, "The League and Evangelism," 20.

41. J. G. Vos, "The Spirit of Error," *The Evangelical Student* 2 (October 1926): 7.

42. "Bulletin of the League of Evangelical Students," September 1930, President's Office Papers, Accession 2006-18, L. S. Chafer correspondence, Box 13, File 16, Archives, Dallas Theological Seminary.

43. Lewis Sperry Chafer to C. Stacey Woods, January 31, 1938, President's Office Papers, Accession 2006-18, L. S. Chafer correspondence, Box 16, File Woods, Archives, Dallas Theological Seminary.

44. W. H. Griffith Thomas, *God Hath Spoken* (Philadelphia: Bible Conference Committee, 1919), 13.

45. Leander S. Keyser published his semiannual textbooks lists in *CFSC*. See, for example, October–December 1923: 63; July–September 1924: 13; and in *Christian Fundamentalist*, July 1927: 20–21; June 1928: 8. Gray suggested a list of textbooks in "Editorial Notes," *Moody Monthly* (December 1923): 141. *Moody Monthly* also carried periodic news about the Association of Conservative Colleges (ACC). See, for example, August 1923: 563; November 1923: 109; January 1925: 220. The ACC published its safe list in the *Christian Fundamentalist* (November 1927): 32.

46. William Bell Riley, "Editorial," *Christian Fundamentalist* (December 1927): 31.

47. Russell, "Thomas Todhunter Shields, Canadian Fundamentalist," 274; George S. May, "Des Moines University and Dr. T. T. Shields," *Iowa Journal of History* (July 1956): 197–99.

48. May, "Des Moines University," 202–7.

49. William Bell Riley, "The Redemption of Des Moines University," *Christian Fundamentalist* (August 1927): 6.

50. J. Frank Norris, "A Modern Miracle," *Fundamentalist* (June 24, 1927): 8–9; "Educated-What for?" *King's Business* (May 1928): 269–70.

51. Russell, "Thomas Todhunter Shields," 274–75; May, "Des Moines University," 211–18.

52. May, "Des Moines University," 223.

53. Russell, "Thomas Todhunter Shields," 276; May, "Des Moines University," 224–25.

54. Russell, "Thomas Todhunter Shields," 276; May, "Des Moines University," 226–27; 230–31; *Christian Fundamentalist* 3 (July 1929): 269.

55. No fundamentalist publication carried news of Shields's humiliation, in contrast to their earlier defense of Shields in Toronto. Only Riley's *Christian Fundamentalist* carried even a brief notice without comment that Wayman had taken a new pastorate (July 1929: 269).

56. Bryan to J. R. Kester, 31 January 1923, Bryan papers, Library of Congress, Washington, DC.

57. LaDonna Robinson Olson, *Legacy of Faith: The Story of Bryan College* (Hayesville, NC: Schoettle Publishing Co., 1995), 49–50.

58. Olson, *Legacy of Faith*, 147–48.

59. William Bell Riley, "William Jennings Bryan University," *CFSC* 7 (October–December 1925): 52.

60. William Bell Riley, "A Fundamentalist University in Chicago," *CFSC* 8 (January–March 1926): 21.

61. "Bryan's Benediction," *King's Business* (December 1925): 534–35; Olson, *Legacy of Faith*, 4.
62. Riley, "William Jennings Bryan University," 52; Riley, "A Fundamentalist University in Chicago," 21; Darien Austin Straw, "A Christian University," *CFSC* 8 (January–March 1926): 17.
63. Virginia L. Brereton, *Training God's Army: The American Bible School, 1880–1940* (Bloomington, IN: Indiana University Press, 1990), 83.
64. J. Frank Norris, "Great rejoicing! 100% of Baylor Faculty sign creedal statement which is 100% for Fundamentalism," *Searchlight* (November 28, 1924): 1.
65. Riley, "The Redemption of Des Moines University," 6.
66. *Moody Monthly* 29 (February 1929): 273.
67. Charles A. Blanchard, "Questionnaire for possible teachers" and "Fundamentals of the Christian Faith," April 15, 1923? Blanchard papers, Wheaton College Archives, Buswell Library; "Student Recruitment," Box 2, College-related file.
68. Charles A. Blanchard, "An American Christian University," *CFSC* 8 (January–March 1926): 15.
69. Trollinger, *God's Empire*, 83.
70. James M. Gray, "Editorial Notes," *Moody Monthly* 21 (April 1921): 347.
71. James M. Gray, "Bible Institutes and Theological Seminaries," *Moody Monthly* 21 (April 1921): 349–51.
72. Curtis Lee Laws, "Fundamentalism from the Baptist Viewpoint," *Moody Monthly* 23 (September 1922): 15.
73. Olson, *Legacy of Faith*, 4, 6, 34.
74. Bob Jones Sr., *Bob Jones Magazine* 1 (July 1929): 1.
75. Mark Taylor Dalhouse, *An Island in the Lake of Fire: Bob Jones University, Fundamentalism, and the Separatist Movement* (Athens, GA: University of Georgia Press, 1996), 47.
76. Ibid., 36.
77. Ibid., 39. See also Daniel L. Turner, *Standing without Apology: The History of Bob Jones University* (Greenville, SC: Bob Jones University Press, 1997), 27–28.
78. Turner, *Standing without Apology*, 25.
79. Ibid., 23–24.
80. Ibid., 20.
81. Bob Jones Sr., *The Perils of America, or, Where Are We Headed?* (n.p., n.d.), 35.
82. Ibid., 13.
83. Ibid., 19.
84. Bob Jones Sr., "Worse than a Common Thief," *Bob Jones Magazine* 1 (June 1929): 1.
85. *Bob Jones College Annual Bulletin*, 3 (April 1929), Bob Jones University archive; Bob Jones Sr., *Original Intentions of the Founder* (Greenville, SC: Bob Jones University, 1960), Bob Jones University archive.
86. Bob Jones Sr., *Bob Jones Magazine* 1 (June 1928): 3.
87. Dalhouse, *Island in the Lake of Fire*, 40–41.
88. Jones, *Perils of America*, 35.
89. Turner, *Standing without Apology*, 30.

90. Bob Jones Sr., "WHY . . . Bob Jones University was founded; WHY . . . It has made so many world-wide contacts in so short a time; WHY . . . it does not hold membership in a regional educational association" (n.p., 1949?) Bob Jones University archives.
91. Turner, *Standing without Apology*, 30.
92. Ibid., 36. See also Dalhouse, *Island in the Lake of Fire*, 44.
93. Turner, *Standing without Apology*, 35, 57; Dalhouse, *Island in the Lake of Fire*, 42.
94. *Bob Jones Catalogue*, 1931–32, Bob Jones University archives.

Chapter 8

1. *Atlanta Constitution*, June 20, 1925, 7.
2. Ward W. Keesecker, *Legal Status of Bible Reading and Religious Instruction in Public Schools* (Washington, DC: United States Government Printing Office, 1930), 2.
3. *Atlanta Constitution*, June 20, 1925, 7.
4. Ibid.
5. Lloyd P. Jorgenson, *The State and the Non-Public School, 1825–1925* (Columbia, MO: University of Missouri Press, 1987). See also David B. Tyack, "Onward Christian Soldiers: Religion in the American Common School," in *History and Education: The Educational Uses of the Past*, ed. Paul Nash (New York: Random House, 1970), 212–55; Joan DelFatorre, *The Fourth R: Conflicts over Religion in America's Public Schools* (New Haven, CT: Yale University Press, 2004), 12–66.
6. James C. Carper and Thomas C. Hunt, *The Dissenting Tradition in American Education* (New York: Peter Lang, 2007), 121–57; Carl F. Kaestle, *Pillars of the Republic: Common Schools and American Society, 1780–1860* (New York: Hill & Wang, 1983), 93, 98–99.
7. Horace Mann, "Letter to Mr. Smith," in *The Bible, the Rod, and Religion in Common Schools*, Matthew Hale Smith (Boston: Redding & Co., 1847), 24.
8. See Donald E. Boles, *The Bible, Religion, and the Public Schools* (Ames, IA: Iowa State University Press, 1963); Keesecker, *Legal Status of Bible Reading and Religious Instruction in Public Schools*; R. Freeman Butts, *The American Tradition in Religion and Education* (Boston: Beacon Press, 1950); J. E. Wood, "Religion and Public Education in Historical Perspective," *Journal of Church and State* 14 (1972): 397–414. The following states (with the year the law passed) required daily Bible reading: Massachusetts (1826), Pennsylvania (1913), Tennessee (1915), New Jersey (1916), Alabama (1919), Georgia (1921), Delaware (1923), Maine (1923), Kentucky (1924), Florida (1925), Idaho (1925), Arkansas (1930). In Indiana, Iowa, Kansas, North Dakota, and Oklahoma, Bible reading was permitted but not required. In Colorado, Connecticut, Maryland, Michigan, Minnesota, Missouri, New Hampshire, New Mexico, North Carolina, Oregon, Rhode Island, South Carolina, Texas, Utah, Vermont, Virginia, West Virginia, Wyoming, and New York Bible reading was allowed in practice. In California, Wisconsin, Illinois, Arizona, Louisiana, Mississippi, Montana, Nebraska, Nevada, Ohio, South Dakota, and Washington, the reading of the Bible in public schools was not allowed, due either to state court decisions, state attorney general decisions, state superintendent of education decisions, or simply to the "general consent of implied prohibition." See Keesecker, *Legal Status of Bible Reading and Religious Instruction in Public Schools*, 3.

9. In many places, the reading of the Bible in public school continued long after the Supreme Court's decision. See Kenneth M. Dolbeare and Phillip E. Hammond, *The School Prayer Decisions: From Court Policy to Local Practice* (Chicago, IL: University of Chicago Press, 1971).

10. Clarence Benson, *Moody Bible Institute Monthly* [*Moody Monthly*] 26 (January 1926): 223–24.

11. William Jennings Bryan, "The Menace of Darwinism," *The Commoner* 21 (April 1921): 5–8.

12. Keesecker, *Legal Status of Bible Reading and Religious Instruction in Public Schools*, 2.

13. Florida House Bill 86, *Journal of the House of Representatives of the State of Florida 1923*, 122, 384.

14. "Ohio Governor Vetoes School Bible Bill," *New York Times*, May 1, 1925, 3.

15. "Advances School Bible-Reading Bill," *New York Times*, January 3, 1926, E20.

16. West Virginia House Bills 176 and 345, *Journal of the House of Delegates of the State of West Virginia 1923*, 70, 136.

17. West Virginia House Bill 1, *Journal of the House of Delegates of West Virginia 1927*, 18, 115, 374, 430, 578, 479, 496, 510, 548, 550; West Virginia House Bill 63, *Journal of the House of Delegates of West Virginia 1927*, 36.

18. West Virginia House Bill 93, *Journal of the House of Delegates of West Virginia 1927*, 41, 390, 431, 499, 555, 588, 591, 787, 813–14, 842, 852, 858–60, 946, 955–58, 988, 996–97, 1082.

19. *Journal of the House of Delegates of West Virginia 1927*, 65, 97–98; West Virginia House Bill 264, *Journal of the House of Delegates of West Virginia 1927*, 104, 663.

20. Harbor Allen, "Supreme Kingdom's Campaign," *CFSC* 8 (October–December 1926): 53.

21. David M. Hovde, "Sea Colportage: The Loan Library System of the American Seamen's Friend Society, 1859–1967" *Libraries & Culture* 29 (1994): 409.

22. Arno C. Gaebelein, "Editorial Notes," *Our Hope* 34 (September 1927): 143.

23. Paul C. Gutjahr, *An American Bible: A History of the Good Book in the United States, 1777–1880* (Stanford: Stanford University Press, 1999), 32–33; Peter J. Wosh, *Spreading the Word: The Bible Business in Nineteenth-Century America* (Ithaca, NY: Cornell University Press, 1994); Kenneth Scott Latourette, "The American Bible Society: A Century and a Half of Global Adventure," *Religion in Life* 35 (1966): 450–56.

24. Michael H. Harris and Gerard Spiegler, "The Fear of Societal Instability as the Motivation for the Founding of the Boston Public Library," *Libri* 24 (1927): 249–75.

25. Boles, *The Bible, Religion, and the Public Schools*, 263.

26. Leonard J. Moore, *Citizen Klansman: The Ku Klux Klan in Indiana, 1921–1928* (Chapel Hill, NC: University of North Carolina Press, 1991), 41–43.

27. Quoted in Kenneth T. Jackson, *The Ku Klux Klan in the City, 1915–1930* (Chicago, IL: Elephant Paperbacks, 1992), 205.

28. 'An Exalted Cyclops of the Order,' "Principles and Purposes of the Knights of the Ku Klux Klan," in Hiram W. Evans, et al., *Papers Read at the Meeting of the Grand Dragons Knights of the Ku Klux Klan, Together with other articles of Interest to Klansmen* (New York: Arno Press, 1977), 128.

29. John Galen Locke, "A Klansman's Obligation as a Patriot to his God, his Country, his Home, and his Fellowmen," *Papers Read at the Meeting of Grand Dragons*, 61.

30. Kathleen Blee, *Women of the Klan: Racism and Gender in the 1920s* (Berkeley: University of California Press, 1991), 145.

31. Christopher N. Cocoltchos, "The Invisible Empire and the Search for the Orderly Community: The Ku Klux Klan in Anaheim, California," in *The Invisible Empire in the West: Toward a New Historical Appraisal of the Ku Klux Klan of the 1920s*, ed. Shawn Lay, 104 (Chicago, IL: University of Illinois Press, 1992).

32. Jackson, *Klan in the Cities*, 166.

33. Blee, *Women of the Klan*, 143–44.

34. W. C. Howells, "Donahey Defies Klan in Vetoing Bible Measure," *Cleveland Plain Dealer*, May 1, 1925, 1, 3.

35. J. W. Northrup, "The Little Red Schoolhouse is One of the Most Sacred of American Institutions," *Imperial Nighthawk* 1 (22 August, 1923): 3.

36. W. S. Fleming. *God in Our Public Schools*, 3rd ed. (1942; repr., Pittsburgh: National Reform Association, 1947), 17.

37. Ibid., 90.

38. Ibid., 80–85.

39. Ibid., 217.

40. S. Charles Bolton and Cal Ledbetter Jr., "Compulsory Bible Reading in Arkansas and the Culture of Southern Fundamentalism," *Social Science Quarterly* 64 (September 1983): 670–76.

41. Pat Ledbetter, "Texas Fundamentalism: *Secular Phases of a Religious Conflict, 1920–1929*," *Red River Valley Historical Review* 6 (Fall 1981): 38–52. Italics in the original title.

42. Willard B. Gatewood Jr., *Preachers, Pedagogues and Politicians: The Evolution Controversy in North Carolina, 1920–1927* (Chapel Hill, NC: University of North Carolina Press, 1966), 101–5.

43. Fleming, *God in Our Public Schools*, 219, 222.

44. George M. Marsden, *Fundamentalism and American Culture, The Shaping of Twentieth-Century Evangelicalism, 1870–1925* (New York: Oxford University Press, 1980), 159.

45. M. H. Duncan, *Modern Education at the Cross-Roads* (Chicago, IL: Bible Institute Colportage Association, 1925), 32.

46. William Jennings Bryan to John A. Taylor, April 17, 1925, Bryan Papers, Library of Congress, Washington, DC.

47. William Jennings Bryan, "Bible Instruction in Schools," May 1, 1925, mimeographed press release in Bryan Papers.

48. T. C. Horton, "Making the Country Safe for the Children," *King's Business* (December 1920): 1111. Emphasis in original.

49. Cortland Myers, "The Crime of Our Godless Schools," *King's Business* 15 (May 1924): 270.

50. John Murdock MacInnis, "We Are Debtors to the Boys and Girls of America," *King's Business* 17 (April 1926): 188.

51. S. M. Ellis, "Secularized Public Schools—The Nation's Menace," *King's Business* 15 (October 1924): 623, 624.

52. T. C. Horton, "Perplexing Problems Confronting our Country," *King's Business* 15 (August 1924): 478.

53. D. C. Smith, "Problems in Christian Education," *Moody Monthly* 30 (January 1930): 234.

54. Johannes G. Vos, "The School-Bag Gospel League," *Moody Monthly* 27 (February 1927): 289.

55. "The Million Testaments Campaign for Students," *Moody Monthly* 31 (June 1931): 501; Rollin T. Chafer, "Editorial Comment: Superiority Complex of Teachers of Naturalism," *Evangelical Theological College Bulletin* 7 (June 1931): 4.

56. Elizabeth Morrell Evans, interview by Robert Shuster, October 8, 1984, collection 279, transcript, Billy Graham Center Archives.

57. Ibid. See also Margaret Lamberts Bendroth, *Fundamentalism and Gender, 1875 to the Present* (New Haven, CT: Yale University Press, 1993), 88; and Bendroth, *Fundamentalists in the City: Conflict and Division in Boston's Churches, 1885–1950* (New York: Oxford University Press, 2005), 168.

58. Bertram J. Youde, ed., *The Biolan 1928* (Los Angeles, CA: Bible Institute, 1928), 91; "Fourth Annual Nuntius Track Meet and Picnic," *King's Business* 19 (July 1928): 423.

59. "The Euodia Conference," *King's Business* 15 (July 1924): 419; Eugene Riddle, ed., *The Biolan 1930* (Los Angeles, CA: The Associated Students of the Bible Institute of Los Angeles, 1930), 44–46. See also Bendroth, *Fundamentalism and Gender*, 84.

60. Sophie Shaw Meader, "The Children's Garden," *King's Business* 15 (July 1924): 429; 15 (August 1924): 495; 15 (September 1924): 567.

61. Edgar McAllister, ed., *The Biolan 1927* (Los Angeles, CA: Bible Institute, 1927), 56, 58.

62. Earl F. Morgan, ed., *Biola Alumni Annual* (Los Angeles, CA: Bible Institute, 1923), 42.

63. Quoted in Kenneth Taylor [Head of the Colportage Department, 1948–1957], "Gold Behind the Ranges," *Christian Life* (June 1948): 26, clipping in the Moody Literature Mission File, Moody Bible Institute archive. For an expanded study of the mission to Appalachia, see Adam Laats "The Quiet Crusade: The Moody Bible Institute and the Mainstreaming of Appalachia, 1921–1966," *Church History* 75 (September 2006): 565–93.

64. *Preaching the Gospel in Print* (Chicago, IL: Bible Institute Colportage Association, 1921).

65. *29th Annual Report of the D. L. Moody Missionary Book Funds* (Chicago, IL: Bible Institute Colportage Association, 1924). Emphasis in original.

66. *Moody Monthly* (September 1934) back cover. Emphasis in original.

67. *Where Hungry Souls Await the Bread of Life* (Chicago, IL: Bible Institute Colportage Association, n.d.). Emphasis in original.

68. *'Holding Forth The Word of Life . . .' to THOUSANDS in Army Camps, Prisons, Hospitals, Mountain and Pioneer Districts . . . through the PRINTED PAGE* (Chicago, IL: Bible Institute Colportage Association, n.d.).

69. *The Bible Institute Colportage Association* (Chicago, IL: Bible Institute Colportage Association, 1896).

70. *Preaching the Gospel in Print.*

71. Arline Harris, "Free Print for the Hungry," typewritten report, Moody Literature Mission file, Moody Bible Institute archive, 1949, 9; "The Plan of Working," in *Where Hungry Souls Await the Bread of Life* (Chicago, IL: Bible Institute Colportage Department, n.d.); Gene Getz, *MBI: The Story of the Moody Bible Institute* (Chicago, IL: Moody Press, 1969), 247.

72. Clarence Benson, "Our Monthly Potpourri," *Moody Monthly* 29 (January 1929): 247.

73. T. C. Horton, ed., *The Gospel of John* (Chicago, IL: Bible Institute Colportage Association, 1922), 69, 79.

74. My calculations are based on several sources. The most complete source was the monthly reports of Book Fund performance, published between 1921 and 1938 in the pages of the *Moody Monthly*. Even these reports, however, were incomplete, since the Book Fund reports were occasionally omitted from crowded issues of the magazine. Another useful source was the file of annual reports of the Book Funds. These contained yearly totals for cash donations and literature deliveries. The archival file of these reports, however, is incomplete. One further problem with the computation of total numbers of books delivered to public schools is that there was no accurate record kept of school deliveries. Although the vast majority of the "mountain" books went to public schools, not all of them did. The records contain occasional hints about the ratio between total book deliveries and those intended for schools, and I based my estimate on this ratio. For example, the August 1929 monthly report contained a note that 454 out of 479 deliveries to the "Mountain" fund were made to public school teachers. In light of all these approximations and estimations, I used the lowest possible number to calculate the totals of books received. By this reckoning, it seems very likely that Appalachian public schools received *at least* 760,906 books between 1921 and 1930. This does not include the number of tracts delivered, but it does include all other categories of book. A total of 802,807 was multiplied by the 1929 ratio to determine this number of books going to public schools alone.

Chapter 9

1. H. L. Mencken, *Prejudices: Fifth Series* (New York: Alfred A. Knopf, 1926), 111.

2. John Roach Straton, quoted in "Hails Rising Spiritual Tide," *New York Times*, August 27, 1925, 12.

3. Gerald Winrod, quoted in "Start World Fight Against Evolution," *New York Times*, May 9, 1927, 4.

4. J. Frank Norris, *The Searchlight* 4 (January 13, 1922): 1.

5. Bob Jones Sr., *The Perils of America, or, Where Are We Headed?* (n.p., n.d.), 35.

6. Willard B. Gatewood Jr., *Preachers, Pedagogues and Politicians: The Evolution Controversy in North Carolina, 1920–1927* (Chapel Hill, NC: University of North Carolina Press, 1966), 193.

7. Ibid., 154.

8. Ibid., 167.

9. T. T. Martin, *Hell and the High School: Christ or Evolution, Which?* (Kansas City, MO: Western Baptist Publishing Co., 1923), 10, 11.

10. William E. Dever [Mayor of Chicago] to Bryan, 2 July 1923, Bryan Papers, Library of Congress, Washington, DC.

11. William Bell Riley, "The Truth and Teacher Agencies," *Christian Fundamentals in School and Church* [*CFSC*] 8 (January–March 1926): 55.

12. "Action of Educators Indorsed [*sic*] by Kingdom," *Atlanta Constitution*, February 26, 1926, 1.

13. Jones, *Perils of America*, 13, 16.

14. Frank Gaebelein, "Book Reviews," *Our Hope* 32 (May 1926): 64.

15. J. Frank Norris, *Searchlight*, 3 (May 12, 1921): 1.

16. Gerald Winrod in Leo P. Ribuffo, *The Old Christian Right: The Protestant Far Right from the Great Depression to the Cold War* (Philadelphia: Temple University Press, 1983), 92–93.

17. John Roach Straton in "Hails Rising Spiritual Tide," 12.

18. J. E. Conant, "Can Northern Baptists Stay Together?" *Searchlight* 4 (April 4, 1922): 22.

19. William V. Trollinger, *God's Empire: William Bell Riley and Midwestern Fundamentalism* (Madison, WI: University of Wisconsin Press, 1990), 43; *The Searchlight* became *The Fundamentalist* on April 15, 1927, which became *The Baptist Fundamentalist of Texas* on April 29, 1927; which became *The Fundamentalist of Texas* on June 1, 1928.

20. Norris proudly published reprints of Northern newspaper articles about his nicknames in his own magazine. See *Searchlight* 4 (January 6, 1922); *Searchlight* 7 (December 14, 1923): 2.

21. J. Frank Norris, "The National Free Thought (Infidel) Weekly," *Searchlight*, 7 (January 4, 1924): 3.

22. J. Frank Norris, "Pentecostal Preacher 'Fleeces' Flock," *Searchlight* 7 (May 2, 1924).

23. Barry Hankins, *God's Rascal: J. Frank Norris & the Beginnings of Southern Fundamentalism* (Lexington, KY: University Press of Kentucky, 1996), 44.

24. Ibid., 28.

25. J. Frank Norris, "World's Baptist Alliance Repudiates Fundamentals," *Searchlight* 6 (August 3, 1923): 1.

26. J. Frank Norris, "A War against Modernism," *Searchlight* 7 (December 14, 1923): 1.

27. J. Frank Norris, "Clergy open Fight to Oust Modernists; 1,000 Fundamentalists at Calvary Baptist Church 'Unsheath Sword of Living Word Against Wolves,'" *Searchlight* 7 (December 14, 1923): 1.

28. C. Allyn Russell, "Thomas Todhunter Shields, Canadian Fundamentalist," *Ontario History* (December 1978): 270.

29. J. Frank Norris, *Searchlight* 9 (December 4, 1925): 1; J. Frank Norris, *Searchlight* 9 (January 29, 1926): 1; T. T. Shields, "Shall We Continue to Feed the Tiger?" *Searchlight* 10 (March 25, 1927): 1; T. T. Shields, "McMaster Exemplifies the Policy of Antichrist," *The Fundamentalist* (April 15, 1927): 1; J. Frank Norris, "A Modern Miracle," *The Fundamentalist* (June 24, 1927): 1.

30. J. Frank Norris, *The Fence Rail* (April 23, 1917): 1.

31. See, for example, *Searchlight* 7 (March 14, 1924): 3; *Searchlight* (November 7, 1924): 3.

32. J. Frank Norris, "Judge Wilson, K. C's, Ku Klux Klan and Bootleggers," *Searchlight* 4 (May 12, 1922): 1.

33. J. Frank Norris, *Searchlight* 7 (November 7, 1924): 3.

34. J. Frank Norris, "Roman Catholicism Versus Protestantism," *Searchlight* 5 (July 14, 1922): 1.

35. J. Frank Norris, *The Fundamentalist* 10 (July 8, 1927):1.

36. "J. Frank Norris Kills Ft. Worth Lumber Dealer in Church; Pleads Self Defense," *Waco News-Tribune*, July 18, 1926. See also "Pastor Faces Trial in Fort Worth Today," *New York Times*, November 1, 1926, 23; see also Hankins, *God's Rascal*, 118–20.

37. Bruce Tarrant, "Minnesota: Modern or Mediaeval?" *The Independent* 118, January 1, 1927, 8–9, 28.

38. "Riley Assails 'U' as Fostering State Atheism," *Minneapolis Tribune*, repr. in *CFSC* (April–June 1926): 35.

39. Ibid., 36.

40. *St. Paul Pioneer Press,* March 10, 1927, in Willard B. Gatewood Jr., ed., *Controversy in the 'Twenties: Fundamentalism, Modernism, and Evolution* (Nashville: Vanderbilt University Press, 1969), 309.

41. Ferenc M. Szasz, "William B. Riley and the Fight Against Teaching of Evolution in Minnesota," *Minnesota History* (Spring 1969): 210.

42. William Bell Riley, "The Faith of the Fundamentalists," in Gatewood, *Controversy in the 'Twenties*, 75.

43. Harbor Allen, "'Supreme Kingdom's' Campaign," *CFSC* 8 (October–December 1926): 51–53.

44. William Bell Riley, *CFSC* 8 (January–March 1926): 28.

45. William Bell Riley to Bryan, 3 May 1923, Bryan Papers.

46. William Bell Riley, *CFSC* 4 (July–September 1922): 5.

47. William Bell Riley, "Funny Fundamentalists," *CFSC* 8 (April–June 1926): 43.

48. Riley, "The Faith of the Fundamentalists," 76.

49. Trollinger, *God's Empire*, 43.

50. Ibid., 108–33; 157, 159.

51. Jones, *Perils of America*, 35; Daniel L. Turner, *Standing without Apology: The History of Bob Jones University* (Greenville, SC: Bob Jones University Press, 1997), 23–24; Mark Taylor Dalhouse, *An Island in the Lake of Fire: Bob Jones University, Fundamentalism, and the Separatist Movement* (Athens, GA: University of Georgia Press, 1996), 41.

52. Jones, *Perils of America*, 39.

53. Ibid., 7.

54. Turner, *Standing without Apology*, 21.

55. Michael Kazin, *A Godly Hero: The Life of William Jennings Bryan* (New York: Anchor Books, 2007), 132, 272.

56. Bob Jones Sr. to Melton Wright, 4 December 1957, Bob Jones University archive; Bob Jones, "WHY . . . Bob Jones University was founded; WHY . . . It has made so many world-wide contacts in so short a time; WHY . . . it does not hold membership in a regional educational association" (n.p., 1949?), Bob Jones University archive.

57. W. E. Patterson, "Intelligence and Orthodoxy," *Bob Jones Magazine* 1 (October 1928): back cover.

58. *Bob Jones College Annual Bulletin* 3 (April 1929).

59. Bob Jones Sr., "Christian Education," transcript of a speech at Dubose Bible Conference, February 1, 1929, Bob Jones University archives.

60. William Martin, "The Transformation of Fundamentalism between the World Wars," in *Critical Moments in Religious History*, ed. Kenneth Keulman, 154 (Macon, GA: Mercer University Press, 1993).

61. "The Bible Crusader's [*sic*] Challenge" *CFSC* 8 (April–June 1926): 56.

62. Ronald L. Numbers, *The Creationists: From Scientific Creationism to Intelligent Design* (Cambridge, MA: Harvard University Press, 2006), 72–76.

63. Arthur I. Brown, *The Antievolution Works of Arthur I. Brown*, ed. Ronald L. Numbers (New York: Garland Publishing, 1995), 9.

64. Ibid., 133.

65. Ibid., 9.

66. Ibid., 155.

67. Numbers, *The Creationists* (2006), 76–87.

68. Harry Rimmer, "Modern Science and the Youth of Today" (Los Angeles, CA: Research Science Bureau, 1925), in *The Antievolution Pamphlets of Harry Rimmer*, ed. Edward B. Davis, 461 (New York: Garland Publishing, 1995); Numbers, *Creationists* (2006), 78–79.

69. Numbers, *Creationists* (2006), 108–11, quote on 108.

70. G. M. Price to Bryan, 1 July 1925, Bryan Papers.

71. Joel A. Carpenter, *Revive Us Again: The Reawakening of American Fundamentalism* (New York: Oxford University Press, 1997), 3.

72. Ronald L. Numbers, ed., *Creation-Evolution Debates* (New York: Garland Publishing, 1995), 160–61.

73. Ibid., 186; Numbers, *Creationists* (1993), 142.

74. Ronald L. Numbers, ed., *Early Creationist Journals* (New York: Garland Publishing, 1995), ix–x.

75. Numbers, *Creationists* (1993), 114–19, 137–60.

Chapter 10

1. Thomas J. Gillespie, "We Are Proud of the Names We Are Called," *Moody Bible Institute Monthly* [*Moody Monthly*] 28 (January 1928): 215. Emphasis in original.

2. Carl F. H. Henry, *The Uneasy Conscience of Modern Fundamentalism* (Grand Rapids, MI: Eerdmans Publishing Co., 1947), xx.

3. Joel A. Carpenter, *Revive Us Again: The Reawakening of American Fundamentalism* (New York: Oxford University Press, 1997), 191–209.

4. James M. Gray, "Editorial Notes," *Moody Monthly* 21 (October 1920): 54.

5. Ferenc M. Szasz, "Three Fundamentalist Leaders: The Roles of William Bell Riley, John Roach Straton, and William Jennings Bryan in the Fundamentalist-Modernist Controversy" (PhD dissertation, University of Rochester, 1969), 213, 70.

6. William Vance Trollinger, Jr., *God's Empire: William Bell Riley and Midwestern Fundamentalism* (Madison, WI: University of Wisconsin Press, 1990), 63.

7. Ibid., 65.

8. George W. McPherson, *The Crisis in Church and College* (Yonkers, NY: Yonkers Book Co., 1919), 195–96.

9. See Mina Carson, *Settlement Folk: Social Thought and the American Settlement Movement, 1885–1930* (Chicago, IL: University of Chicago Press, 1990); Molly Ladd-Taylor, *Mother-Work: Women, Child Welfare, and the State, 1890–1930* (Champaign, IL: University of Illinois Press, 1995); Anthony M. Platt, *The Child Savers: The Invention of Delinquency*, rev. ed. (Chicago, IL: University of Chicago Press, 1977).

10. Louise Bowen, *Safeguards to City Youth at Work and at Play* (New York: MacMillan, 1914), 5.

11. William Edward Biederwolf, *What About So-Called Christian Evolution?* (Chicago, IL: Bible Institute Colportage Association, 1926), 4.

12. William Jennings Bryan, "The Danger of Reckless Teaching," *The Commoner* 21 (December 1921): 11.

13. "Child-centered or Bible-centered, Which?" *Moody Monthly* 30 (January 1930): 231.

14. Virginia L. Brereton, *Training God's Army: The American Bible School, 1880–1940* (Bloomington, IN: Indiana University Press, 1990), 87–91.

15. M. H. Duncan, *Modern Education at the Cross-Roads* (Chicago, IL: Bible Institute Colportage Association, 1925), 26, 27.

16. Elizabeth Morrell Evans, interview by Robert Shuster, October 8, 1984, collection 279, transcript, Billy Graham Center Archives.

17. Susan Strasser, *Satisfaction Guaranteed: The Making of the American Mass Market* (New York: Pantheon Books, 1989), 99.

18. William Jennings Bryan, *The Bible and Its Enemies: An Address Delivered at the Moody Bible Institute of Chicago* (Chicago, IL: Bible Institute Colportage Association, 1921), 42.

19. Charlie Oakes in Michael Kazin, *A Godly Hero: The Life of William Jennings Bryan* (New York: Anchor Books, 2007), 300.

20. Alfred Fairhurst, *Atheism in Our Universities* (Cincinnati, OH: Standard Publishing Co., 1923), 120.

21. Curtis Lee Laws, *Christian Fundamentals in School and Church* [*CFSC*] 7 (October–December 1925): 66–67.

22. Jeffrey P. Moran, *The Scopes Trial: A Brief History with Documents* (Boston: Bedford/ St. Martin's Press, 2002), 62; Raymond D. Boisvert, *John Dewey: Rethinking Our Time* (Albany: State University of New York [SUNY] Press, 1998).

23. Robert H. Wiebe, *The Search for Order, 1877–1920* (New York: Hill and Wang, 1967), 208. Wiebe described modernist Protestant theologians as "the honorary chairmen of progressivism."

24. *Evangelical Theological College: First Annual Announcement, 1924–1925* (Dallas, TX: Evangelical Theological College, 1925), 5; John D. Hannah, "The Social and Intellectual Origins of the Evangelical Theological College" (PhD dissertation, University of Texas at Dallas, 1988), 116–21, 148–95. Quotation is on page 159.

25. Lewis Sperry Chafer to the Sunday School Literature Committee, 18 October 1922, President's Office Papers, Accession 2006-18, box 15, file 13, Archives, Dallas Theological Seminary.

26. A. B. Winchester to Lewis Sperry Chafer, 23 April 1923, President's Office Papers, Unprocessed Chafer Papers, Accession 2006-22, box 1, file "Correspondence Concerning Founding," Archives, Dallas Theological Seminary.

27. Lewis Sperry Chafer to A. C. Gaebelein, 13 November 1923, President's Office Papers, Accession 2006-18, box 12, file 23, Archives, Dallas Theological Seminary.

28. A. C. Gaebelein to William Bell Riley, 9 December 1922, President's Office Papers, Unprocessed Chafer Papers, Accession 2006-22, box 1, file "Sunday School Literature Committee," Archives, Dallas Theological Seminary.

29. Lewis Sperry Chafer to J. Frank Norris, 13 November 1923, President's Office Papers, Accession 2006-18, box 13, file 30, Archives, Dallas Theological Seminary; Lewis Sperry Chafer to A. C. Gaebelein, 13 November 1923; Lewis Sperry Chafer, "Careless Misstatements of Vital Truth," *Our Hope* 30 (March 1924): 540–51.

30. Lewis Sperry Chafer to Ralph D. Smith, 26 December 1925, President's Office Papers, Accession 2006-18, box 14, file "Smith, Ralph D. and Bible House of Latin America," Archives, Dallas Theological Seminary.

31. Ibid.

32. Lewis Sperry Chafer to the Sunday School Literature Committee, 18 October 1922.

33. Lewis Sperry Chafer to A. C. Gaebelein, 13 November 1923.

34. Lewis Sperry Chafer to Oliver Buswell, 19 February 1930, President's Office Papers, Accession 2006-18, box 10, file 26, Archives, Dallas Theological Seminary.

35. C. G. Trumbull to Lewis Sperry Chafer, 13 November 1922, President's Office Papers, Accession 2006-18, box 15, file 13, Archives, Dallas Theological Seminary.

36. C. G. Trumbull to James M. Gray and Lewis Sperry Chafer, 9 December 1922, President's Office Papers, Accession 2006-18, box 15, file 13, Archives, Dallas Theological Seminary.

37. Lewis Sperry Chafer, "Effective Ministerial Training," *Evangelical Theological College Bulletin* 1 (May 1925): 7.

38. *Evangelical Theological College Bulletin* 1 (January 1925): 3.

39. Lewis Sperry Chafer to the Sunday School Literature Committee, 18 October 1922.

40. Lewis Sperry Chafer to Oliver Buswell, 3 July 1929, President's Office Papers, Accession 2006-18, box 10, file 26, Archives, Dallas Theological Seminary.

41. *Evangelical Theological College Bulletin* 2 (November 1925): 6.

42. Rollin T. Chafer, "Editorial Comment: Spirit-Directed Intellectuality," *Evangelical Theological College Bulletin* 4 (May 1928): 3.

43. Rollin T. Chafer, "Editorial Comment: Literary Bunk and Hokum Exposed," *Evangelical Theological College Bulletin* 5 (November 1928): 2.

44. Rollin T. Chafer, "Superiority Complex of Teachers of Naturalism," *Evangelical Theological College Bulletin* 7 (June 1931): 2–4.

45. Lewis Sperry Chafer to Robert Dick Wilson, 20 December 1923, President's Office Papers, Accession 2006-18, box 16, file 23, Archives, Dallas Theological Seminary.

46. A. C. Gaebelein to Lewis Sperry Chafer, 14 April 1925, President's Office Papers, Accession 2006-18, box 12, file 23, Archives, Dallas Theological Seminary.

47. Lewis Sperry Chafer to Oliver Buswell, 19 February 1930.

48. George M. Marsden, *Fundamentalism and American Culture, The Shaping of Twentieth-Century Evangelicalism, 1870–1925* (New York: Oxford University Press, 1980), 174.

49. J. Gresham Machen, *Christianity and Liberalism* (Grand Rapids, MI: Eerdmans Publishing Co., 1923), 14–15.

50. J. Gresham Machen and Charles P. Fagnani, "Does Fundamentalism Obstruct Social Progress?" *Survey Graphic* 5 (July 1924): 391; Machen to R. S. Kellerman, 7 October 1924, J. Gresham Machen Papers, Montgomery Library archive, Westminster Theological Seminary, Philadelphia; Machen to John Holliday Latane, 5 November 1928, Machen papers.

51. J. Gresham Machen, "What Fundamentalism Now Stands For," *New York Times*, June 21, 1925, XX1.

52. D. G. Hart, "J. Gresham Machen, Confessional Presbyterianism, and the History of Twentieth-Century Protestantism," in *Re-Forming the Center: American Protestantism, 1900 to the Present*, ed. Douglas Jacobsen and William Vance Trollinger Jr. (Grand Rapids, MI: Eerdmans Publishing Co., 1998), 129.

53. Frederick Schweitzer to Machen, 18 June 1930, Machen papers.

54. Machen to Andrew S. Layman, 6 August 1929, Machen papers.

55. Kendrick C. Hill to Machen, 27 March 1928, Machen papers.

56. Machen to Kendrick C. Hill, 14 March 1928, Machen papers.

57. Frederick Erdman to Leander S. Keyser, 24 March 1928, Machen papers.

58. James M. Gray, "Editor's Note," *Moody Monthly* 28 (January 1928): 215.

59. James M. Gray, "Scholarship and Evangelical Christianity," *Moody Monthly* 29 (October 1928): 53.

60. See, for example, James M. Gray, "Editorial Notes," *Moody Monthly* 25 (August 1925): 533.

61. James M. Gray, "Editorial Notes" *Moody Monthly* 26 (October 1925): 48.

62. J. C. O'Hair, "Why I Am a Fundamentalist," *Moody Monthly* 27 (August 1927): 576–78.

63. J. H. Ralston, "Our Monthly Potpourri," *Moody Monthly* 26 (May 1926): 428.

64. Clarence H. Benson, "Our Monthly Potpourri," *Moody Monthly* 27 (May 1927): 440.

65. James M. Gray, "Editorial Notes" *Moody Monthly* 26 (October 1925): 47.

66. James M. Gray, "Christianity and False Evolutionism," *Moody Monthly* 26 (February 1926).

67. James M. Gray, "Editorial Notes," *Moody Monthly* 26 (April 1926): 364.

68. James M. Gray, "Editorial Notes," *Moody Monthly* 27 (August 1927): 569.

69. James M. Gray, "Editorial Notes, *Moody Monthly* 27 (June 1927): 472.

Conclusion

1. Maynard Shipley, "Growth of the Anti-Evolution Movement," *Current History* (May 1930): 330.

2. William Adams Brown, "After Fundamentalism—What?" *The North American Review* 223 (September 1, 1926): 406.

3. Gerald Skoog, "The Coverage of Human Evolution in High School Biology Textbooks in the 20th Century and in Current State Science Standards," *Science and*

Education 14 (2005): 398; Ronald L. Numbers, *The Creationists: From Scientific Creationism to Intelligent Design* (Cambridge, MA: Harvard University Press, 2006), 264–65.

4. Jeffrey P. Moran, *The Scopes Trial: A Brief History with Documents* (Boston: Bedford/ St. Martin's Press, 2002), 156.

Epilogue

1. Reginald Stuart, "U.S. Court to Hear Arguments on Creationism," *New York Times* December 7, 1981, A21; Reginald Stuart, "'Creation' Trial: Old South Against New," *New York Times*, December 13, 1981, 38; Marcel C. La Follette, "Creationism in the News: Mass Media Coverage of the Arkansas Trial," in *Creationism, Science, and the Law: The Arkansas Case*, ed. La Follette (Cambridge, MA: MIT Press, 1983), 189–207.

2. Clarence Darrow, "Evolution a Crime," *New York Times*, January 12, 1982, A15; and H. L. Mencken, "Scopes: Infidel," *New York Times*, January 12, 1982, A15.

3. Duane Gish, *Teaching Creation Science in Public Schools* (El Cajon, CA: Institute for Creation Research, 1995), v. See also Jerry Bergman, *The Criterion: Religious Discrimination in America* (Richfield, MN: Onesimus Press, 1984), v.

4. Ronald L. Numbers, *Darwinism Comes to America* (Cambridge, MA: Harvard University Press, 1998), 91.

5. The Gallup Organization, "Public Favorable to Creationism," *Gallup.com*, February 14, 2001, http://www.gallup.com/poll/2014/public-favorable-creationism .aspx (accessed December 17, 2007); Ronald L. Numbers, *The Creationists: From Scientific Creationism to Intelligent Design* (Cambridge, MA: Harvard University Press, 2006), 1.

6. Paul F. Parsons, *Inside America's Christian Schools* (Macon, GA: Mercer University Press, 1987), x.

7. Alan Peshkin, *God's Choice: The Total World of a Fundamentalist Christian School* (Chicago, IL: University of Chicago Press, 1986), 26.

8. *Characteristics of Private Schools in the United States: Results from the 2001–2002 Private School Universe Survey*, NCES 2005-305, U. S. Department of Education (Washington, DC: National Center for Education Statistics), 9.

9. See Adam Laats, "Inside Out: Christian Day Schools and the Transformation of Conservative Protestant Educational Activism, 1962–1990," in *Inequity in Education: A Historical Perspective*, ed. Debra Meyers and Burke Miller (Lexington, KY: Rowman and Littlefield Press, 2009), 183–209.

10. Phillip Smith, personal interview, June 5, 2008; Walter Fremont, "The Christian School Movement Today," speech delivered July 31, 1989, audiotape in Bob Jones University Archives; A. A. Baker, *The Successful Christian School: Foundational Principles for Starting and Operating a Successful Christian School* (Pensacola, FL: A Beka Book Publications, 1979). See also Adam Laats, "Forging a Fundamentalist 'One Best System:' Struggles over Curriculum and Educational Philosophy for Christian Day Schools, 1970–1989," *History of Education Quarterly*, 50 (February 2010).

11. Gary Coombs, "ACE, An Individualized Approach to Christian Education," *Interest* (September 1978): 9–10; Gary Coombs, "History and Development of ACE,"

CLA Defender 1 (1978): 6, 25–26; Accelerated Christian Education, *Facts about Accelerated Christian Education* (Lewisville, TX: Accelerated Christian Education, n.d.); Walter Fremont, "The Christian School Movement Today," audiotape of lecture given July 31, 1989, in Bob Jones University Archive.

12. Dan B. Fleming and Thomas C. Hunt, "The World as Seen by Students in ACE Schools," *Phi Delta Kappan* (March 1987): 523.

13. Ronald E. Johnson, "ACE Responds," *Phi Delta Kappan* (March 1987): 520–21.

14. Gerald Skoog, "The Coverage of Human Evolution in High School Biology Textbooks in the 20th Century and in Current State Science Standards," *Science and Education* 14 (2005): 398, 404–5.

15. Lloyd P. Jorgenson, *The State and the Non-Public School, 1825–1925* (Columbia, MO: University of Missouri Press, 1987), 135; R. Freeman Butts, *The American Tradition in Religion and Education* (Boston: Beacon Press, 1950), 192.

16. R. B. Dierenfield, *Religion in American Public Schools* (Washington, DC: Public Affairs Press, 1962), 51.

17. Dorothy Nelkin, *Science Textbook Controversies and the Politics of Equal Time* (Cambridge, MA: MIT Press, 1977), 27–30; Skoog, "Coverage of Human Evolution," 405.

18. Numbers, *The Creationists* (2006), 277.

19. Sean Cavanagh, "'Intelligent Design' Goes on Trial in Pa.," *Education Week* 25 (October 5, 2005): 1, 16–17.

20. Numbers, *The Creationists* (2006), 218–19.

21. *Abington School Dist. v. Schempp*, 374 U.S. 203 (1963); Donald E. Boles, *The Bible, Religion, and the Public Schools* (Ames, IA: Iowa State University Press, 1963), 131–44.

22. Kenneth M. Dolbeare and Phillip E. Hammond, *The School Prayer Decisions: From Court Policy to Local Practice* (Chicago, IL: University of Chicago Press, 1971), x, 28.

23. "Report: The month's Worldwide News in brief," *Moody Monthly* 64 (January 1964): 8.

24. "The Supreme Court Speaks," *Moody Literature Mission News* 4 (1963), Moody Literature Mission [MLM] File, Moody Bible Institute [MBI] Archive. See also Adam Laats, "The Quiet Crusade: Moody Bible Institute's Outreach to Public Schools and the Mainstreaming of Appalachia, 1921–1966," *Church History* 75 (September 2006): 565–93.

25. Joel A. Carpenter, *Revive Us Again: The Reawakening of American Fundamentalism* (New York: Oxford University Press, 1997), 3.

26. Jonathan Zimmerman, *Whose America? Culture Wars in the Public Schools* (Cambridge, MA: Harvard University Press, 2002), 131–211.

27. *An Encouraging Report*, Colportage department fundraising brochure, [1948?] MLM File, MBI Archive.

28. . . . *and Many Believed: A Report from Colportage Department of Moody Bible Institute*, Colportage department fundraising brochure, [1954?] MLM file, MBI Archive; *MLM News*, February 1960; *MLM News*, December 1961; *MLM News*, October 1962; *MLM News* no. 2, 1966; *An Encouraging Report*.

Index